His Brother's Keeper

ALSO BY JONATHAN WEINER

Time, Love, Memory

The Beak of the Finch

The Next One Hundred Years

Planet Earth

His Brother's Keeper

{A STORY FROM THE EDGE OF MEDICINE}

Jonathan Weiner

Fourth Estate • *London* and *New York*

First published in Great Britain in 2004 by
Fourth Estate
A Division of HarperCollins*Publishers*
77–88 Fulham Palace Road
London W6 8JB
www.4thestate.com

1

Portions of this book, in different form, first appeared in *The New Yorker*.

A catalogue record for this book is available from the British Library.

ISBN 0–00–714218–8

Printed in Great Britain by
Clays Ltd, St Ives plc

For my mother

If you would like to know what men really are, the time to learn comes when they stand in danger or in doubt. For then at last words of truth are drawn from the depths of the heart, and the mask is torn off, reality remains.

LUCRETIUS,
The Way Things Are

{ CONTENTS }

{ AUTHOR'S NOTE }

I have followed the story of the Heywood brothers since the first months of 1999, when Jamie transformed himself from a mechanical engineer into a genetic engineer and began his race to save Stephen. In this book, where I reconstruct what was spoken in the past or quote someone else's reconstruction, I put the passage in italics.

Part One

The Key in the Door

For every birth a death.

Lucretius

{ ONE }

Portents

When they were boys, Jamie and Stephen Heywood loved to arm wrestle. They made it a ritual: first the right arm, then the left, then, if there was time, a wrestling match on the rug. Their rules of engagement were so complicated and so long unspoken that no one else ever learned the game. Even Jamie's best friend Duncan Moss did not know how to play. Duncan would take one step across the line on the rug. Then he would see the look on Jamie's face.

What? What did I do?

He did not know the rules.

The Heywoods lived in an old house on Mill Street in Newtonville, Massachusetts, a suburb of Boston. All three of the Heywood boys, Jamie, then Stephen, and then the youngest, Ben, were athletic dreamers, inventors of many rituals and adventures. The house on Mill Street is a block from a patch of woods and a pond. The brothers played football there in a corner field by the house of a neighbor they called Aunt Betsy. Late at night when it rained hard, Jamie and Stephen snuck out with boogie boards. They hopped a fence to the creek, which got roaring in a good storm. Through the dark and the rain they rode the rapids out of Bullough's Pond.

Their parents, Peggy and John Heywood, are well known in

Newtonville. They love traditions, too. Each of them has served terms as Senior Warden of Grace Church, in Newton Corner. When their boys were young, they went back every summer to the dairy farm in South Dakota where Peggy had grown up. She had won a full scholarship to Radcliffe, in Cambridge, which is where she met John. Peggy worked as a therapist; she kept her practice small and devoted herself to the family.

Every seven years, they spent a year in England, where John was born and raised. John Heywood is a professor of mechanical engineering at the Massachusetts Institute of Technology, and an international authority on internal combustion engines. He is the son of a British coal engineer who turned to solar power early on, back when only the cranks were interested in what he had to say—back when a maverick who worried about coal smoke, soot, and acid, and praised the power of the sun, was like a bolt from the future. John Heywood consults about energy efficiency for Ford in Detroit, Ferrari in Italy, Toyota in Japan. When he is at home, he runs MIT's Sloan Automotive Laboratory, to which he commutes from Mill Street on a bicycle.

Each summer in July or August, Peggy's side of the family gathered from across the country for a reunion at the beach town of Duck, near Kitty Hawk, in the Outer Banks of North Carolina. A few of the children on the beach were always honorary Heywoods, as Aunt Betsy was an honorary aunt. John and Peggy, Jamie, Stephen, and Ben each brought friends. The boys swam, sailed, surfed, and raced on the cold shining track the waves made for them. Every summer on one of their last days at Duck they played a game of basketball at the beach, wickedly violent games that routinely sent one of their cousins to the hospital.

As Jamie and Stephen got older, they kept arm wrestling, ritually. In their late teens and early twenties, when the two and a half years between them no longer mattered, they were perfectly matched. But Jamie became a mechanical engineer, like their father and his father in England. Jamie was intelligent and driven, and spent his days and nights working at a desk. Stephen became a carpenter, a hands-on man like their mother's father and brother in South Dakota. Stephen was intelligent, too, but he mistrusted desks and ambitions. He spent a few years swinging a hammer on a framing crew, and his right arm became unbeatable.

Late in July of 1997, when Jamie was thirty and Stephen was twenty-eight, they arm wrestled in the beach house their parents had rented that summer at Duck. Jamie was five feet, eleven and three-quarter inches tall, and he weighed one hundred and sixty-five pounds; Stephen was six-foot-three, two-twenty. Jamie was keeping himself in shape, but Stephen was building his first house that year, and his right bicep and tricep were very well-defined. In arm wrestling, there is always a moment when the winner knows he has won and the loser knows he has lost. The brothers were both surprised when they realized that for the first time in five years, Jamie would force Stephen's right arm down to the table.

Jamie whooped. *I beat my carpenter brother. I'm the man! I'm the man!*

Stephen won the next bout, which they fought, as always, left-handed. That shut up Jamie.

Neither of them suspected that anything was wrong.

That year a team of scientists and veterinarians in Scotland announced the birth of a strange lamb, the identical twin of its mother. The news hung above the year like a comet. All around the world, the arrival of the lamb was received as a portent, like an earthquake, a fire, an eruption, a millennial battle won or lost. Something was out of whack in the order of the world and would have to be put right, if it could ever be put right—or else turned to advantage, transformed into acts of healing as novel as the conception of that cloned lamb.

That was also the year the world's front pages carried the story of the death of Jeanne Louise Calment, from Arles, France. She helped inspire people to hope that in the new millennium, human beings might live as long as Methuselah. Jeanne Louise Calment was 122 years old. She remembered Vincent van Gogh.

Those who loved science and those who mistrusted it felt an almost supernatural touch of hope or dread that year, as if all our human rituals were about to change forever. Some scientists and doctors dreamed of a new, regenerative medicine. Soon, maybe as soon as the turn of the millennium, they might be able to control the forces of generation and regeneration that bring us all into being and maintain our bodies as long as we are here, powers that had labored unseen on the face of the earth since the first quickening of life. Lucretius invokes those powers at the start of *The Way Things Are,* his epic poem in praise of what we now call science, composed in Rome before what we now call the first millennium. "Mother of Love!" Lucretius begins, calling on the goddess Venus, "to whom all living creatures owe their birth." Mother of Love, for whom "the artful earth sends up sweet flowers." Mother of Love, who made "the first rise of things," who created "the seeds of all things," the goddess who makes the birds and beasts seek their own kind in the spring, and so renews the world.

"O Mother of Love, help me write my verses!" the poet pleads, opening the story of science with a prayer.

Now miracles seemed possible, likely, imminent: the curing of the incurable, the repair of the irreparable, the saving of lives that in any other generation would have been given up for lost. But the cures might come or they might never come. They might come that year or not for another hundred years; and nobody knew. So when a man, woman, or child contracted a disease—one of the innumerable diseases that were still fatal and incurable—the doctor, the patient, the patient's family and friends all wondered if they were forced to accept defeat or if they could fight and win.

Under the sign of the comet, a certain kind of medical story entered the news more and more often. Families in need of miracles were moved to acts of extraordinary courage. Here and there a man or a woman, like a hero in an ancient legend, ran to the edge of medicine and a little over the edge, racing into the science of the future, storming the walls and setting them on fire. These people committed acts of selfishness and great selflessness, of self-interest and the purest altruism. They were trying to save one life, and they were trying to save the world.

I am a science writer. I spent the last years of the twentieth century looking over the shoulders of biologists who watched evolution in action. I went to the Galápagos Islands, also called Darwin's Islands. The beaks of their finches and mockingbirds helped lead Darwin to his theory of evolution by natural selection, and his followers can now see Darwin's process at work in real time as it sculpts and resculpts the beaks of his finches. I also visited many of the molecular biologists whose laboratories are now, for better and for worse, our new Galápagos.

Being a science writer means being a perpetual student, and those trips gave me some of the greatest lessons of my life. I will never forget the moment when one of Darwin's mockingbirds perched at the top of my spiral notebook, as bold as a bird in the Garden of Eden, to watch my pen point squiggle on the page. In the laboratories, mutant fruit flies also landed on my notebook, which was a less glamorous experience. But out in the field or inside the laboratories, I thought I had come as close as anyone could ever come to the process that Darwin discovered at the heart of life.

By 1997, I knew that this work was coming closer to home than the Galápagos or the islands and archipelagoes of the world's universities. That year Darwin's process was a matter of human experiment and experience. This is what we seemed to see everywhere at the edge of medicine—evolution, the potential for evolutionary change, the genetic engineering of human beings. So I decided to tell a story that would take me out of the field and the lab, and into human experience.

To write about medicine has always been to confront the whole human experience. "Medicine takes you there," as the director of a famous research hospital told me later, when I met him in the university cafeteria and he heard what I was doing on campus. "Medicine *demands* that you are there." This is more true in our time than in any since Hippocrates, as the field of medicine moves toward its new power over first and last things, over birth and death, beginnings and endings. Beginnings and endings have always been the times to learn what people really are. Around a birth or a death there are moments when you seem to see the entire span of a human life, and sometimes even the span of the species—the whole story, beginning, middle, and end, from first to last.

And now even more than before, when the edge of medicine may change human nature forever.

I followed many stories at the turn of the millennium, but one of them came closest to home for me. It was the story of two brothers, an engineer and a carpenter: a brilliant young mechanical engineer who turned himself almost overnight into a genetic engineer in his race to save the carpenter. For a little while I became an honorary Heywood, too, or wanted to be one.

The Heywoods' story taught me many things about the nature of healing in the new millennium. It also taught me about what has not changed since the time of the ancients and may never change as long as there are human beings—about what Lucretius calls "the ever-living wound of love."

The Heywoods mean the whole story to me now, an allegory from the edge of medicine. A story to make us ask ourselves questions that we have to ask and do not want to ask. How much of life can we engineer? Are there any lines we cannot or should not cross?

What would you do to save your brother's life?

{ TWO }

The Family Artist

In snapshots from one of their first summers on the farm in South Dakota, Jamie sits in his uncle's old go-cart, making it *go.* Stephen squints off into the sky, holding a butterfly net. As a child, Stephen was always wandering off with that butterfly net and getting lost. He ran ahead at the Statue of Liberty and the National Air and Space Museum and got lost. Once on a family camping trip in the Scottish moors he wandered right out over the horizon. He had to be rescued by Heywood search parties.

Stephen was the biggest reader of the three brothers, and he might have been the most intellectual, but he did not like the pretentiousness of intellectuals. He thought about them and studied them in a sort of askance way, but he was determinedly not one of them. Even lovable and unpretentious academics like his father worked on an abstract plane that floated somewhere above the way things are. It is still a joke today with the Heywood boys that their father never worked on the family cars.

"I did change the oil once or twice," John says.

"The author of *Internal Combustion Engine Fundamentals* has never worked on a car except to change the oil," says Jamie.

"Nobody does," says John. "People who know the theory don't know the practice. I don't know practice. If I bring my car in, I make it a *practice* not to tell them my diagnosis, because I'm so very often wrong."

John Heywood is wiry, with a graying, auburn, close-clipped British mustache. His right eye angles toward his nose, but his left eye has a level and candid gaze. In the magnifying lenses of his glasses, his wandering eye used to terrify the boys' friends when they were small. He works sixty-hour weeks at the lab. Peggy is just as wiry, and in spite of the sophistications of life around Boston you can see at a glance the farmer's hardworking daughter. Neither John nor Peggy is tall, and from early on their sons towered over them.

Jamie and Stephen both loved to read the kind of science-fiction novel in which a boy growing up on a farm or a suburban planet discovers that he is destiny's child, born to save the galaxy. One of their favorite books, *Ender's Game* by Orson Scott Card, begins when Ender Wiggin, a small, brilliant boy of six, is yanked from his family to be raised in a space station. Ender and his friends in space believe they are just playing training games, computerized war games. In reality they are blasting real enemy starships, slaughtering the Buggers and saving the human race.

Jamie would not even take Orson Scott Card's books into the bathroom. That would have been sacrilegious. But when Stephen was reading a paperback he liked, he tore it down the spine and gave Jamie the part he had already read.

In London, during John's first sabbatical leave from MIT, he did some consulting work in Coventry and drove back to his family in a loaned Jaguar. Until that point the Heywoods had never owned anything better than a secondhand Ford. They were all seduced by the experience. They drove around London with the Jag's sunroof open, blaring the old hymn that John Heywood loved to sing in his tolerable baritone in church choirs: *Bring me my Arrows of Desire . . . Bring me my Chariot of Fire.*

John and Peggy sent Jamie to one of the better schools in London. Stephen and Ben ended up in a local state school instead, where Stephen machined a hammer out of steel. He wrote a funny essay about the contrast between the students. Jamie's friends were all diplomat's sons, spoke forty languages, and walked as if they already ruled the world. Stephen's friends were all in gold chains and looked like pimps.

In his teens, Jamie tinkered with bigger and better engines and the software of the Heywoods' first IBM computer. At night in the woods behind the house on Mill Street he and a half-dozen of his friends held epic battles of laser tag. Some of them remember now that Jamie took those battles much more seriously than Stephen, or anyone else in the game. One of Jamie's friends at Newton North High School called him King of the Geeks. Stephen felt comfortable with all kinds of people, and he ran with many crowds: the toughs, the Beautiful Ones, the artists, the actors. The Heywoods' cousins called Stephen the Male Model. He was the best-looking of all the cousins, and there were a lot of them. Girls followed him.

This is a first son's dream: *I am going to be just like my father.* A second son says, *I will be different.* The first son says, *I am going to be just like Dad but better—cosmically better—and save the world.*

Both Jamie and later Ben went through their father's department at MIT and graduated with degrees in mechanical engineering. But like many middle sons, Stephen rejected his father's line of work and went looking for one of his own. He went to Colgate, where he experimented with writing, painting, and what he later decided was more than his share of illegal substances. He wandered after college, too, tooling across the country on a customized Harley-Davidson in a brown leather motorcycle jacket and brown leather boots, a diamond in one ear. With his girlfriend, Stephanie—they had been together on and off since they were sixteen—Stephen backpacked in the Rocky Mountains. Wherever he lived, he picked up odd jobs painting and carpentering.

Stephen was now as sturdy as he was tall, with thick black hair, a clear, steady gaze, and a strong, square, cleft chin. He wore T-shirts and khakis and a three-day stubble.

Jamie's forehead was higher and his cleft chin tapered to a point.

He had the kind of bright pale face that signals health as much as good color. His high-wattage shine, along with his intensity, and his remarkably regular features, made him look like a man from the future. He became something of a clotheshorse. He liked expensive Eastern-establishment suits. Sometimes he dressed like the young Turks and techno-prophets: black Italian jackets with dark shirts and dark silk ties that set off his pale face and had their own mystic and expensive sheen.

Stephen was happy to be a twentysomething in the generation that called itself X, for expectations unknown. He was ironic, he was laconic, and he did not mind saying he was lazy. He enjoyed being a slacker.

"I've been sponging off my parents for years," he used to say. "I'm totally a Gen Xer."

"Stephen totally defines it," said Jamie.

But the Heywood brothers stayed close. They were all exceptionally talented, and they had an old-fashioned style of family pride. When Ben graduated from high school, Stephen sent him a small check and a scrawled note:

> What, a letter from Steve? This is for graduation, and also to tell you that I'm proud of you. You have managed to do a heck of a lot better than either James or I did in school, and I respect that. I also think that you have gotten what you deserve by getting into MIT. You should continue to do what you have done in the past, as I think you have something that 1 & 2 may be missing. Always remember, be a Heywood.

Eventually Stephen and his girlfriend landed in an apartment in San Francisco. Stephen put up a few notes on bulletin boards in hardware stores and found part-time jobs painting garages and building decks. After six months, he and Stephanie broke up.

By Alamo Square Park, there is a row of Victorians, the famous Painted Ladies, one of the only rows of fine old houses that survived the great San Francisco earthquake and fire of 1906. Stephen used to sit in the park and study the Painted Ladies. One day an eccentric architect, Mark Little, saw a card of Stephen's in a hardware store. The architect built and restored fine houses with something like Victorian high style. He hired Stephen, first part-time, then full-time, and became his mentor; and eventually Stephen decided to restore a house on his own.

In 1996, he found a dilapidated cottage in the Crescent Park neighborhood of Palo Alto, which is just south of San Francisco. Stephen studied the neighborhood from his Harley. Palo Alto reminded him of Newtonville: palm trees instead of maples, but the same professional people, the same engineers, psychotherapists, professors on bicycles. He thought he understood real estate in Palo Alto, and the cottage was just what he was looking for. It was tiny, and it was a wreck, a junker, but it was surrounded by homes that were worth easily half a million, a million, and more.

Stephen was broke. His credit card debt was $15,000. But the other Heywoods helped. His brother Ben had just split up with his girlfriend, too—his fiancée. Ben quit his job at a bioengineering company, cashed in his stock options, and went partners on the cottage with Stephen. John and Peggy put up most of the down payment—calling it an investment. The brothers bought their Heartbreak Hotel for $325,000, which approached what their parents' house was worth back in Newtonville.

After the signing, Stephen rode straight to the cottage on his Harley. He and Ben had gotten the place on a low bid because they had waived their right to an inspection. The owner was an old woman who had let it rot around her and then abandoned it. The tiny lot was overgrown with tall weeds, the stucco was mottled and crumbling, the asphalt roof leaked, the walls were stained, the halls were packed with junk, and half the wiring was dead. Stephen had to bushwhack to the front door, crawl through the junk in the hallways on his hands and knees, and plug one of the feed wires into the meter before he could turn on a light.

The brothers camped out in the place and hired half a dozen college dropouts and Ivy Leaguers at loose ends to share it and work with them. Stephen learned to draw blueprints, to pour foundations, to flirt with the planning lady at City Hall. He bought a black Ford pickup truck. He and Ben and their crew knocked down so many walls that they had to move out. After a year or two there was nothing much left of the cottage. Of the original frame, Stephen had saved only one beam, and whatever relics he had preserved were all shining, polished, varnished, and standing in new places. When the city inspector came by to see the work in progress, he crossed his arms, shook his head, annoyed and amused. *I thought this was supposed to be a renovation job.*

Stephen loved the project, right down to the blueprints. One of his cousins, David Searls, visited Palo Alto in the middle of the construction work. He remembers Stephen standing over a set of complicated plans. "It just struck me that it might have been a work of art to him," he says. "It was sort of that pose." The pose of the artist lost in his labors.

{ THREE }

"Anything Is Possible Now"

From the beginning it has been a great dream of Western Civilization that science can save. "Examining the body requires sight, hearing, smell, touch, taste, and reason," Hippocrates said, about two thousand four hundred years ago. This is the first note of what is distinctive in Western medical tradition, a note you do not find expressed in any systematic way in the traditions of the mummy-wrappers of Egypt or the shamans of Tibet. It is part of a larger faith that reason can help us; and of course medicine is the most intimate and profound way that reason can help us. But how long it took to get much better than shamans through the use of reason.

One of the founding dates in science is the year 1543, when Copernicus's *De Revolutionibus* (On the Revolutions) and Vesalius's *De Humani Corporis Fabrica Libri Septum* (Seven Books on the Structure of the Human Body) appeared in print just a few weeks apart. But it was a long time until scientists could send any satellites revolving in orbit, and it was a long time before anatomists could do much more than anatomize. Vesalius's beautiful charts did not cure anybody, they only showed readers how to perform more dissections. When Vesalius himself did cure the head wound of a sick prince, people thought he had practiced witchcraft, and he was forced to make a pilgrimage to the Holy Land. On the way home, his ship was wrecked on the island of Zante, near the Peloponnesus—named by the ancient Greeks for Pelops, a symbol of resurrection. Vesalius died on Zante.

A Spanish doctor named Michael Servetus argued for the circulation of the blood in 1553, but (for his many heresies) he was burned at the stake, like Giordano Bruno, who argued for life on a multitude of worlds.

The long rise of science that began in the Renaissance did little at first to change the prognoses and the lives of the sick. The best doctors knew this, and were humble. During a siege of Turin, the first military campaign of the great Renaissance surgeon Ambroise Paré, part of the castle was captured by one Captain le Rat. "He received an arquebus-shot in his right ankle, and fell to the ground at once, and then said, 'Now they have got the Rat,' " wrote Paré. "I dressed him, and God healed him." That was the surgeon's refrain in the 1500s: "I dressed him, and God healed him."

A court physician to King James proved in 1628 that the blood does circulate and the heart is a pump. William Harvey gave confidence to generations thereafter that the body could be studied as a mechanism: Its organs worked like pumps, valves, filters, clocks. Our bodies are mechanisms that can be understood. But more than three hundred years had to pass before surgeons could act on that discovery of Harvey's and fix the valves of the pump.

Likewise, naturalists looking through microscopes could see germs as long ago as the early 1700s. Their descriptions of what they saw excited people's imaginations. Daniel Defoe in *Journal of the Plague Year* writes that plague "might be distinguished by the Party's breathing upon a piece of Glass, where the Breath condensing, there might living Creatures be seen by a Microscope of strange, monstrous and frightful Shapes, such as Dragons, Snakes, Serpents and Devils, horrible to behold." But it was not until the late 1800s that Robert Koch in Germany and Louis Pasteur in France established the germ theory of disease. And it was not until the discovery of antibiotics in the mid-twentieth century that doctors finally had powerful multipurpose weapons against infections.

In January 1901, in New York City, a small boy died of scarlet fever at the age of three. His name was John Rockefeller McCormick, and he was the first grandchild of John D. Rockefeller. Even for an heir of the richest man in the history of the world, doctors in 1901 could do nothing but sit up with the family while they prayed, and help them try to keep their composure. The one-word motto of the great nineteenth-century doctor Sir William Osler, the single word he offered as a proverb to graduating medical students, was *aequanimitas*. And *aequanimitas,* the equanimity of the ancients, was still almost all that doctors had to offer in 1901, not only for scarlet fever but also for typhoid fever, pneumonia, polio, tuberculosis, and flu. Doctors had nothing better in their black bags for diphtheria, dysentery, syphilis, or strep throat—which is how scarlet fever usually starts. Virtually every disease that had ever visited human beings in any millennium was still incurable.

Rockefeller hired a board, and the board bought the last thirteen acres of open land on the Upper East Side of Manhattan, between 64th Street and 68th Street, on a low cliff above the East River. Soon there was a stone gate at 66th Street and what is now York Avenue, and through the gate a drive lined with sycamores and wrought iron gas lamps, and at the head of the drive a tall somber laboratory, now called Founder's Hall. There, for the first time in their lives, a few scientists could work on the problems of life full-time. This was the first experiment of its kind in America, and there were few like it anywhere else, besides the Pasteur Institute in Paris and the Koch Institute in Berlin. One of the celebrated German biologists who was lured to join the Rockefeller Institute wondered how he could find enough to do at a laboratory bench to fill a whole day.

By hiring the best scientists he could find and letting them work in this cloister at the edge of Manhattan, without obliging them to take on either students or patients and protecting them from responsibilities or distractions of any kind, Rockefeller hoped to help transform medicine. And he did. But it was not until 1944 that a few scientists working in the Rockefeller research hospital discovered that genes are made of DNA, work that led to the discovery of the double helix in Cambridge, England, in 1953, and the rise of genetic engineering.

By the last decade of the twentieth century, the power of human understanding had reached a point at which it was possible to view all of life as a project in molecular engineering, and the saving of life as nothing more than engineering and genetic carpentry. Now surgeons dreamed of operating on molecules inside the bodies of their patients—hammering, sawing, and polishing, like Stephen in his Heartbreak Hotel. They dreamed of teaching the body to heal and repair itself molecule by molecule. The molecular surgeons had a new and exalted maxim. To teach the body to heal itself molecule by molecule would be to heal like God.

The double helix became the greatest icon of science, and the new symbol and totem of medicine, where it replaced the ancient twining snakes of the caduceus. Molecular biologists started companies and made millions. A few dreamed of fortunes bigger than Rockefeller's. But others admitted how little they knew. Most medical practice was still empirical. That is, doctors' success with patients owed at least as much to experience, trial and error, as it did to scientific systems and theories. While knowledge of molecules grew explosively, the number of lifesaving drugs grew very, very slowly. Twenty-four centuries after Hippocrates, and after a century of biomedicine—of healing with the help of modern science—most drugs worked without anyone knowing why.

The question became: Now that scientists could anatomize at the level of genes and molecules, now that they could engineer there, how long until new cures?

Hope asked: How soon can we use this invisible anatomy to repair a dying nerve, brain, or heart?

Fear asked: How much of the body can we change without losing the patient we hoped to save? Will science change human nature? Will we change the human heart?

In the spring of 1997, while Stephen's house was taking shape, the news of the birth of the cloned lamb Dolly seemed to illuminate the edge of medicine with power, glory, irrational exuberance, and terror. I remember the moment when I first realized how wild the mood was, not only for the world looking on, but for some of the scientists themselves. Early in March 1997, I went to Princeton to hear a lecture by one of the leaders in the field of advanced reproductive technologies, ART—the ART that created Dolly. The lecturer was Jacques Cohen, director of the Institute for Reproductive Medicine and Science of St. Barnabas in Livingston, New Jersey, who was always pressing the outer limits of what could be done at his infertility clinic.

That night as I drove to Princeton, lambs and colts were being born in the barns of Pennsylvania and New Jersey—farm animals conceived the old-fashioned way. In the trees the buds were making their preparations. By then the whole world was talking about cloning. Dolly had seemed to come out of nowhere, out of some secret underworld that now promised or threatened to change life on earth.

The news of Dolly's birth was announced in the United States in the *New York Times* on its front page on Sunday, February 23, 1997: "Scientist Reports First Cloning Ever of Adult Mammal." Near the top of her story, the reporter Gina Kolata quoted a Princeton biologist, Lee Silver, who was just finishing a book about ART; his title was *Remaking Eden.* Silver told Kolata that he would have to tear up his manuscript. "It's unbelievable," he said. "It basically means that there are no limits. It means all of science fiction is true. They said it could never be done and now here it is, done before the year 2000." On March 4, President Clinton had called for a moratorium on the cloning of human beings "until our Bioethics Advisory Commission and our entire nation have had a real chance to understand and debate the profound ethical implications of the latest advances." The chair of the commission was Harold Shapiro, the president of Princeton.

St. Barnabas is not far from Princeton and New York City; but that night it was as if we were meeting a master spy who came in from the cold for one evening from the other side. My friends and I waited for Jacques Cohen in the parking lot behind Lewis Thomas Laboratory, which is the home of molecular biology at Princeton. Cohen arrived in a Porsche and he brought with him the urbanity of a medical man and also the fresh and canny air of business. He was a quiet, contained, bald but youthful-looking man in his late forties. He spoke with a Dutch accent.

We ate in the Lewis Thomas Laboratory, which is a Doge's palace of a building, designed by the architect Robert Venturi as a beautiful boast of the power of the molecular view of life. I sat across the table from Cohen. During dinner he stayed dry and deadpan most of the time. Now and then he flashed a wide, slightly embarrassed grin, sharing the absurdity of life. He confided to me that he had done a lot of work lately and not published it. He also said that a day or two after the headlines about Dolly, he had gotten an offer from one of his patients.

"I've known him for years. The man called and said—I wasn't sure if he was joking—'One million dollars if you will clone me.' "

"What did you say?"

"I told him, 'Add a zero.' "

For a moment Cohen kept his face expressionless. Then he flashed me his absurdist grin. I must have stared at him, because he repeated the story word for word to make sure I understood. "I wasn't sure if he was joking . . ."

In other words, Cohen had spoken as ambiguously as his client, and made what could have been taken as a counteroffer.

Lee Silver sat at the head of the table. He would be sharing the podium with Cohen. Dolly and the *New York Times* had changed his life, he said. Unlike most of his colleagues, he had never felt shy about talking to the press. For years he had kept an album at home in which he pasted clippings from the local papers, the *Princeton Packet,* the *Princeton Weekly Bulletin,* and the *Bergen Record.* Now, since the story in the *New York Times,* he had been on twenty-two radio shows and three television shows. That day alone he had given two or three interviews. He had also buried his uncle and scattered a few spadefuls of earth over the coffin, according to the Jewish tradition. *Dust you are and dust you will be.*

Silver looked exhilarated and exhausted, like a prophet whose apocalypse has finally arrived. That month anyone who could speak boldly and authoritatively about cloning was talked hoarse. The bioethicist Art Caplan, at the University of Pennsylvania, was sometimes on more than one show at the same time: You flipped the channel and there he was again. Before Dolly his Web site had gotten 500 hits a day; now it got 17,000. At the Roslin Institute in Scotland, where Dolly was cloned, every phone rang all day. One scientist hurrying down a hall heard something in a broom closet. He opened the door and there was a phone in there, and it was ringing.

Silver told Jacques Cohen that the cloning of Dolly had shocked and thrilled him. "It's changed my whole philosophical view."

"Well," Cohen replied urbanely, with European modesty, in a dry tone of self-effacement, "we must remember that genes are not everything." He himself had a patient's baby about to be born that very night, any minute now—with help from someone else's egg cytoplasm.

Silver asked a question, Cohen gave a technical answer. I could not follow all of it, but I gathered that Cohen and his group had extracted some cytoplasm from the egg of a fertile woman and injected it into the egg of an infertile woman. Then they had added sperm from the patient's husband. From there everything had gone as it usually does with *in vitro* fertilization (IVF), fertilization in glass. They had implanted the fertilized egg in the woman's womb, and now she was about to have a baby.

The lecture was held in the largest hall of the Lewis Thomas Laboratory. Several hundred students were in the audience, including Princeton's graduating class of molecular biologists.

Jacques Cohen's lecture was a dry, smooth, professional slide show. In 1959, the first successful IVF rabbit. In 1978, the first IVF human being, Louise Brown, a baby conceived in a petri dish. Then in the early 1980s, the cloning of sheep, in secret, by an old friend and colleague of Cohen's, Steen Willadsen. Cohen explained, with an edge to his voice, that Willadsen had developed many ingenious techniques that Ian Wilmut, the man who cloned Dolly, had built on and had now made public at last.

When Cohen was done, he sat down at a table next to the lectern, took a sip of water, and let Lee Silver take over.

If Cohen's lecture was low-key, Silver's was in the highest key I ever heard from a podium. He told the students what he had said at dinner, that cloning had changed his view of life. "From the moment I heard it was done," he said, "my entire conception of science changed." He held up a thick stack of paper: his manuscript, *Remaking Eden*. Then he took the first few pages, tore them up, and dropped the pieces on the floor.

"We thought it would always be necessary for sperm and egg to unite to make an embryo," he said. "We thought that was a law. We were wrong."

He ripped up a few more pages of his manuscript and let them flutter down.

"I wonder now if any biological law can't be broken. I don't think there are any biological laws. There's nothing that can't or won't be done in the future.

"We thought it would be impossible. I really thought it would be impossible—though I never like to use that word."

Rip! Rip! Rip!

"Do you realize what this means? You could take a skin cell, a cell we don't respect very much. Scratch the skin and you kill cells—we don't care much. Right?"

He scratched his bare forearm and followed with his eyes the invisible flurry of a thousand lost cells that fell to the floor toward the first pages of his manuscript.

"But now, every one of those cells I'm killing can form a human life. We respected the fertilized egg partly because we thought that only the fertilized egg has the potential to form a human life. That's no longer true. Any one of those cells has the potential to form a human life."

As an embryo's cells divide and divide, their numbers explode and so does their diversity. They become muscle cells, liver cells, kidney cells, hair cells, skin cells, nerve cells. Until Dolly, biologists had believed that all those mature adult cells were committed to their identities: They could not go back. Each adult cell had built so much stiff molecular scaffolding around its DNA that it could never be young again. But Wilmut and his team in Scotland had plucked the DNA from one skin cell in the udder of a ewe, and that DNA had started over and grown into a lamb. It was a miracle of regeneration, and it would soon help to cast a glow of infinite possibility around the new field of regenerative medicine.

Until the turn of the twentieth century, Silver said, biologists had thought that the workings of inheritance would always be secret from human beings. "With Mendel's laws, for the first time, they knew how it worked. Then they thought we would never understand what a gene was. They thought that was something we would never know. And then in 1953 the molecular structure of a gene was determined: the double helix. Pretty incredible!"

In the early 1970s, scientists shocked themselves again. They discovered how to cut and splice DNA in bacteria. They could become genetic engineers. They had thought they could never do that.

"And cloning—everyone said it was impossible. Me too. I was among them. Absolutely impossible. It was an absolute law of nature, an absolute barrier.

"All knocked down! We can now think of ways of cloning almost anything we can imagine! There isn't anything impossible anymore!"

Silver destroyed a few more pages of his manuscript.

"I mean, this is really incredible! Human beings have been around a long time: five million years. We discovered fire three hundred thousand years ago. We tamed the wolf thirty thousand years ago. One hundred years ago we found the laws of the gene. Fifty years ago we found the molecular structure of genes. Twenty-five years ago we found out how to do genetic engineering. And now with cloning—"

Silver broke off in mid-declamation and dropped his voice into a meadow of calm. "This is sort of a free-form talk," he said softly, and the students laughed. "It's not what I was going to talk about two weeks ago. But the world changed two weeks ago."

He threw the last pages of *Remaking Eden* over his head and let them flutter to the floor.

There was so much talk about the lamb that night that I forgot about the baby. That story was more important than Cohen had let on at dinner. It is now regarded in some circles as the moment when our best hopes and worst fears about genetic engineering began to come true. A few of Cohen's colleagues and many of his critics argue that the baby born that night was genetically modified, a GM baby. If so, then the promise and threat we saw in the portent of the lamb had already arrived. Human beings had begun the genetic engineering of human babies. I did not know this yet. But that night all of us felt that genetic engineering was coming of age, and coming home to us.

When the students and scientists had finished mobbing them, Cohen and Silver walked up the long aisle to the projection booth to collect their slides. For some reason the doctor seemed to have found the professor's performance amusing.

"Anything is possible now," he said drily. "You convinced me."

Silver wondered how Ian Wilmut in Scotland was handling his instant celebrity.

"I think Wilmut is surprised," Cohen said. "To him, it was a small step. And to me, too. But I'll stay away from cloning for a while. I need to make a living."

"Add one zero!" I said.

"You can buy anybody."

{ FOUR }

The Key in the Door

Slowly that year Stephen's house became more a work of art than a house. He and Ben worked longer and longer hours, ran farther and farther behind. They went back to the bank and borrowed more and more money. Stephen suffered night sweats from his credit card debts. Sometimes he loved the place and sometimes he hated it.

I'm not a long-term worker kind of guy.

In the summer of 1997, when Jamie beat him at arm wrestling in the Outer Banks, Stephen assumed that he was just tired.

Five months later, on a morning in Christmas week of 1997, Stephen arrived at the house and put his key in the front door. He was more than six months behind his latest revised deadline, and his to-do list nagged at him whenever he got to work. The yard was mud, and the local sod farms were sold out of sod. The roof was untiled: Thousands of dollars' worth of six-inch ceramic tiles were stacked on it in piles, and the roofer refused to return his calls. The locks in the doors were brass, brand-new, and top-of-the-line, but every one of them was sticking.

That Christmas, his whole family was there to help him finish the house. John had taken a sabbatical leave from MIT and moved to Palo Alto. He was writing a book, *The Two-Stroke Cycle Engine,* and helping out with the carpentry. Peggy had closed her practice and retired in order to make the trip. She helped Stephen by designing the new kitchen and bathrooms and laying out the gardens.

Ben was there almost every day, too. He had not done much of the manual labor. He was on his way to business school at UCLA. He did not want to go back to engineering; he wanted to be a producer in Hollywood. At the house, Ben was management. He wrote in his application to UCLA that his job was dealing with the banks, budgets, expenses, subcontractors, and "focusing my brother's efforts on critical-path tasks." Whenever people praised the house, Ben told them that his brother did all the real work. Ben was afraid of ladders, scared of heights. *I don't have Stephen's hands.*

The only one in the family who had to fly out that Christmas was Jamie. Even though they lived on opposite coasts, Jamie and Stephen were still inseparable. They talked on the phone a few times a day; and as often as they could, Jamie and his wife Melinda flew out from their home in Yardley, Pennsylvania, to help with the house. Jamie was making good money at Advent Design Corporation, a small engineering company north of Philadelphia. When the house was almost done, and his brothers' credit cards were all tapped out, Jamie would come around each weekend and leave a check for ten thousand dollars: "So those were expensive weekends."

The neighbors on Forest Avenue and for a few blocks around in Crescent Park kept strolling by that Christmas to admire Stephen's house. The dilapidated little cottage was now a gem. Stephen let Jamie show them around, and Jamie worked each room like a realtor. Stephen had saved and rehung all of the windows from the original house because he loved rippled old glass. The uprights on the sides of the windows were poplar, custom milled. The windows' headpieces and sills were custom, too. Some of them he had turned into double-hung windows, some into casement windows with little brass knobs. Almost all of the hardware was new, including fancy brass hinges at fifty dollars a pair. Stephen had saved the old doors from the original house and had them rehung by a woodwork company near Palo Alto, with new doors specially milled to match. And handsome as it all was, the place met the building code for earthquakes. "You could have turned that house upside down and shaken it," Jamie says. "I really think you could."

Real estate values were rising all over Palo Alto, which is prime Silicon Valley, and Stephen was beginning to realize that after all the sweat and squalor, after all his beginner's idiotic mistakes, after all his forebodings of disaster, he might actually make a profit. If he could just sell the place without losing money, he saw himself restoring more houses. They stood in a row ahead of him like the Painted Ladies. He knew what he wanted to be now—he knew what he was. He could imagine doing this kind of work for the rest of his life, or at least until he turned forty. Everyone in his family could see the change in him. His drifting days were ending. Stephen's mother thought it was wonderfully appropriate that Christmas when Stephen's Harley was stolen from the garage. To Peggy the theft seemed like a gift or at least a wink from heaven. "It was a miracle! It was a miracle!"

That morning Stephen stood on the front porch fiddling with the key in the front door. He had not yet gotten around to fine-tuning the lock, which a carpenter has to do with every door and jamb. The key refused to turn. But the door was beautiful: He had stained, varnished, and polished it to match the door of the garage. Stephen also liked the look of the doorknob and the plate around the keyhole. He had shined the brass as brightly if it carried the name of his first ship, or the Heywoods' heraldic coat of arms. The front doorknob was new and its mate on the inside was a knob that Stephen had saved from the original door, an antique brass globe impressed with an ornate flower. All gleaming brass: the knobs, the lock, the decorative escutcheon around the keyhole.

Stephen unscrewed the front doorknob. He inspected the latch bolt and the strike plate to make sure the bolt lined up with the hole in the plate. It did. He screwed the knob back in place.

Then, experimentally, he tried the key with his left thumb and forefinger. This time the tumblers in the lock revolved smoothly and the cam threw the bolt. So there was nothing the matter with the key or the lock.

Stephen tried again with his right thumb and forefinger. But he could not turn the key.

Jamie told his family that Christmas that he had decided to quit his job at Advent Design Corporation. He was going into biology.

It was a leap into the dark for Jamie. He had never even taken a biology course—he had always been an engineer. Jamie likes to tell a story. In Newtonville, he says, his eighth-grade teacher, Mr. Malagodi, assigned everyone in class to design and build the strongest possible bridge out of simple classroom materials. Most of Jamie's classmates built a bridge with a ruler, for a span of exactly one foot. Jamie's bridge spanned five feet, with yard-high towers made of T squares. It was a suspension bridge: Strings ran down from the T squares to support a series of rulers under the roadway. Most of the one-foot bridges failed with one or two textbooks placed on them. A few made it to five textbooks. "My bridge was last and everyone wanted to see the test," Jamie says. "In the end, we put eighteen textbooks in the center of the bridge before one of the T squares buckled and the whole thing came tumbling down."

At MIT, John Heywood is well known and respected for his competence, intelligence, hard work, and good sense. Both John and Peggy are solid in-the-box people. But at home and at school, Jamie's inventions were always baroque: dreams of brave new companies, new industries. He founded start-ups that broke down, he made wild Icarian flights out of the box. In an MIT manufacturing class called 2.86, the assignment was to design a yogurt cup. Everyone else made a round cup—as instructed. Jamie made his square.

"It was a nice cup," his mother says. "Jamie tended to get As or Cs. He didn't get Bs."

Friends of his from MIT days say he solved problems not so much by mastery of equations and formulas but by an intuitive sense of what would work, what the right angles were for the trusses on the bridge. He winged his answers without calculating them. He could not always find his way by incremental steps, but sometimes he could sense it, and he was always out of the box.

Even his marriage took him out of the box. He met Melinda in Cambridge when he was at MIT and she was at Wellesley. Melinda grew up in the Pickle Family Circus, where her father was the bandleader. Her mother is a belly dancer in Athens. By the age of two, her parents were divorced, and Melinda danced with her mother for the first time. By the age of seven, she was dancing in smoky Greek tavernas until one or two in the morning. Her mother's stage name is Rhea, and Melinda and her sister Piper worked their way through college and graduate school as belly dancers, the Daughters of Rhea. In Christmas week of 1997, Melinda had just finished her PhD in medieval French literature at the University of Pennsylvania, while Piper worked on a PhD in molecular biology at Johns Hopkins—a connection that Jamie would find useful later on. Melinda also worked as a circus dance artist and juggler. She performed in the Big Top of Circus Flora with a troupe that included Flora the Elephant, Nino the Clown, and the famous Flying Wallendas. In one of the company's acts, Melinda, sequined and slender, caught in the crosslights, played the allegorical figure of Hope.

In his first job after graduation, Jamie helped design the hulls of boats for the winning team of the 1992 America's Cup. Then he embarked on a series of entrepreneurial projects, managing teams of engineers, some of them more than twice his age. He fought in business the way he had fought at basketball, football, tennis, wrestling, arm wrestling, and hockey (he was a bruiser at hockey). "Recklessness and competition are just different ages," he says. Actually, Jamie was still reckless at thirty. He was scary on highways—the fast lane was too slow for him. He also raced boats. ("All I do is race. I don't sail. I get bored in a boat if I'm not racing.") He loved the wild starting line, crouching on a 25,000-pound piece of fiberglass as it thrashed almost out of control. He still wanted to save the world. He also yearned for the best boats, the finest cars, the finest houses, and his sense of destiny visibly ate at him.

"Jamie was always a sort of bumptious young man," Peggy says. "His ambitions never fit his skin. He always wanted to do more or be more than he was actually doing or being."

By 1997, Jamie had been working at Advent Design Corporation for six years as an engineer and a project manager. He had helped make a surgical irrigator, a fiber-optic surgical loupe, and a machine for manufacturing suppositories. He and Melinda lived in a fine old three-story Victorian in Yardley—Stephen had done some beautiful work on it for them. Advent had been founded by two techies from MIT, and to Jamie it felt like a sort of idealized extension of school. He was playing with cool toys, designing and building useful gadgets and robots, and the people there were very bright. But the place was too small for him, and that year he decided it was time to move on. He wanted to own a shop of his own, or else try something completely different.

Flying in and out of Silicon Valley on his visits to Palo Alto made Jamie restless. In those years, the workshops in the valley loomed as tall as the towers of Wall Street, even though most of the buildings that housed the new start-ups and dot-coms were only two stories high, no taller than Stephen's Heartbreak Hotel. From desktops and laptops across the country, Jamie and his techie friends e-mailed each other with stock tips, news about initial public offerings, ideas for start-ups and dot-coms, insanely ambitious business plans—and some of his friends were getting rich. Being young and untried was their biggest asset. The social commentator Michael Lewis was in Silicon Valley then, writing a book called *The New New Thing.* "Having a past actually counted against a company, for a past was a record and a record was a sign of a company's limitations," Lewis says. "You had to show that you were the company not of the present but of the future. The most appealing companies became those in a state of pure possibility."

Out in Palo Alto, John and Peggy knew that real estate prices were riding on a bubble, the Silicon Bubble. Markets in Asia crashed in the fall of 1997, and the older Heywoods prayed in Crescent Park: *Stay up, stock market! Stay up, stock market!* The bubble quivered but it did not burst.

At Advent, almost as soon as Jamie began to look around, a businessman he liked made him an offer: a position in an air-conditioning company. Jamie thought the offer was potentially lucrative but not exciting. A friend of his, David Edelman, called his father, the Nobel Prize–winning biologist Gerald Edelman. Gerald Edelman had spent years at Rockefeller's old institute, now called Rockefeller University. There he had run a sort of biological think tank devoted to studies of the brain. He had moved his think tank to La Jolla, California, where he ran it in the spirit of the old Rockefeller. It was called the Neurosciences Institute.

That year, with Wall Street in love with science and technology, Edelman was looking for someone to handle what is known as technology transfer. He wanted a young entrepreneur to study the new ideas and gadgets that his scientists were inventing, and find ways to bring them into the marketplace. The institute's share of the profits would be plowed back into the research.

People at the institute called Jamie. It was a Wednesday. Jamie told them that he was signing a contract with his new employer on Monday. They told him to get on a plane right away.

The Neurosciences Institute is on Torrey Pines Mesa in La Jolla. The place keeps a low profile and it has a very small sign on the road. The first time I went to visit, my cabdriver drove right past it. In designing the place, Edelman tried to capture the feeling of Rockefeller University, which seems, when you step through the stone gates at 66th Street, like a secluded world. The central marble esplanade was planned by one of the nation's greatest landscape architects, Dan Kiley, to suggest a cloistered courtyard and to inspire the scientists who walk there to reach inward, where other public spaces urge people to reach out. The esplanade is lined with sycamores instead of stone columns because Rockefeller University is devoted to the study of life. Kiley loved the sycamores because "their high, open-branching habit" was "appropriately erudite," he wrote, and contrasted nicely with the "stark habit" of the gingko trees. He wanted the scientists and doctors who strolled there in white lab coats to think "of ancient walled gardens founded upon the notion of paradise on earth."

When Jamie walked into the Neurosciences Institute, he found himself in a postmodern version of that esplanade, an open space full of irregular and angular concrete forms and courtyards. The walls were faced with stainless steel, glass, and a beautiful sandstone studded with fossils. The central plaza was paved with terrazzo concrete and serpentine, and it was ornamented with moving water, bamboo, eucalyptus, Torrey pines, cycads, melaleuca, and native grasses. Edelman had included architectural reviews on the institute's Web site. The critic at the *New York Times* had called it "a magnificent piece of work." The critic for the *San Diego Union-Tribune* had said, "This is as good as architecture gets."

The Neurosciences Institute, like Rockefeller, is a not-for-profit research center. Edelman describes it as a scientific monastery. It is supported strictly by private money, and Edelman keeps its staff small because he wants scientists there to do "high risk–high payoff" research. For that kind of work, he thinks small is best. Rockefeller is tiny, but it has been the incubator or the home of more than twenty Nobel Prize winners besides Edelman, who won the prize for research on the workings of the immune system.

Even back at the Rockefeller, Edelman was concerned about the design of his institute down to the smallest detail. A friend of mine there once told me that Edelman used to agonize over rug swatches. There was something absurd about a Nobel Prize–winning immunologist worrying about the pattern in the carpet. But places like that did not just happen. The attention to detail was important, like the attention to detail in cathedrals. Such a beautiful place sent messages about biomedicine to everyone who entered it. Scientists were building on the work and working in the spirit of those who had lived in the millennia before, and they were making a model of reality for those who would live in the millennia to come; and all this was conveyed by a combination of allusions to the monasteries of the past and the space colonies of the future.

Edelman was trained originally as a concert violinist by a classmate of Jascha Heifetz from Leopold Auer's school in St. Petersburg, before he found his way into medicine and then into research. As a Nobelist, he stands for what he likes to call, with light self-mockery, "the religion of Sweden." In his institute above the Pacific he was still walking Rockefeller's esplanade. The word "esplanade" comes from the Latin *explanare,* to flatten out, explain, make plain. Edelman stood for the Path of the Explainers.

After doing his fundamental work on the immune system, Edelman had isolated molecules called N-CAMs, which help nerve cells adhere and bind to other nerve cells—electrical tape for the wiring of the nervous system. From there, Edelman had turned to the workings of the brain, hoping to solve the mystery of consciousness. Like many other biologists, he thought of the brain as the next great frontier in science. The brain is everything in human experience: our awareness, our feelings, our ability to make and create and speak and understand the thoughts of others; a source of health and sickness; a carpenter's ability to hammer a nail or look at a wreck and see a gem that could be built in its place.

So much that we take for granted can go wrong so quickly with the brain and lead to so many illnesses, from alcoholism and anorexia and Alzheimer's to depression, paralysis, paranoia, schizophrenia, xenophobia. In the United States alone, illnesses that begin in the brain have damaged tens of millions of lives, and every year these diseases end the lives of hundreds of thousands. The cost of treating disorders of the brain in this country is more than six hundred billion dollars a year. No other group of illnesses costs us so much money, to say nothing of the anguish they bring to the afflicted and to their families, which is incalculable.

The U.S. Congress had declared the 1990s "The Decade of the Brain." By the last years of the decade more scientists were working on the brain and the nervous system in the United States and around the world than on any other aspect of the science of health. More was learned about the brain and nerves of the body in those years than in all the rest of human history. Yet in many ways the brain remained as mysterious and difficult to treat as ever, because of its extraordinary complexity. Scientists still could not explain the nature of a thought, a memory, a dream. That is, they did not understand the nature of the mind in health, much less in sickness.

Many scientists hoped that the time was right at last for great advances: for basic understanding and for the curing of the incurable. In biomedicine the focus had turned from the curing of germ-borne diseases to the curing of nerve-death diseases like Alzheimer's and Parkinson's. In the United States today the number of Americans with a debilitating neurological disorder is one in five. As the Neurosciences Institute's Web site points out, that number is going to rise as the number of old people in this country rises. In every country rich enough to benefit from the medical advances of the nineteenth and twentieth centuries, nerve-death diseases may become the biggest health problem of the twenty-first, as more and more of the well-off live long enough to die from them. International pressure to find cures for these diseases is enormous. In the world's wealthiest countries, sometime in the twenty-first century the number of people who suffer from an illness of the brain or the nerves may be one in two. What scientists can do now will affect hundreds of millions of families. It may touch the fate of nations. This is a place where a young scientist can hope to save the world.

Gerald Edelman met with Jamie Heywood for a few hours. Jamie was very impressed by him: a tall man, and highly charismatic.

To Jamie, he seemed almost intimidatingly confident of his personal powers and the power of neuroscience. Edelman had revolutionized the study of the immune system, and now he told Jamie that he and the scientists at his institute were revolutionizing the study of consciousness.

Edelman liked what he saw in Jamie, too. Jamie still looked very young that year, but at the turn of the millennium, especially in the think tanks of California, youth was an infinite asset. Jamie did not know biology; but many of the scientists at the Neurosciences Institute were trying to understand the workings of the brain by experimenting with computers and robots. One team had built a robot called NOMAD, which was slowly learning to wander around a playpen. As director of product management at Advent, Jamie designed and built robots for venture capitalists. He was intelligent and hungry, and he could learn to turn ideas into products for the Neurosciences Institute.

Edelman assigned another neurobiologist, Ralph Greenspan, to show Jamie around and court him over lunch. Greenspan is a more gently charismatic man than Edelman, also tall, also a great talker, but soft spoken, with graying wavy hair. He had worked at Princeton, at New York University, and at the pharmaceutical giant Hoffman-La Roche. Greenspan told Jamie that he had joined the institute because he wanted to pursue extremely edgy studies of the sleeping, waking, and possibly conscious brains of fruit flies. The institute gave him the freedom to do that. Over lunch, he worked on Jamie pretty hard. *You can always go into air-conditioning . . .*

Jamie listened and looked around him. He was impressed by the institute's dining room, where the lunches are served by waiters. Here again Edelman was building on Rockefeller, Greenspan explained. That was one of Rockefeller's traditions from the turn of the twentieth century. Lunches there were events, and dinners in the president's house were often served by waiters in black tie. Scientists sometimes made fun of the place for that, but the most extraordinary things had been achieved at Rockefeller: Nobel Prizes, discoveries that transformed biomedicine.

After Jamie left, Edelman talked over the hire with Greenspan. They decided that Jamie would be a good person to figure out how to package the think tank's ideas and turn them into money. He would be their tech-transfer guy, their money guy. His title would be Director of Technology Transfer.

Jamie himself, when he thought it over, did not see his role as just a money guy. He saw himself as joining a high-powered team in one of the greatest team projects he could imagine.

"I got on a plane on a Thursday evening, interviewed on Friday, was offered the job on Saturday, and accepted it on Sunday," Jamie says. "It was less money and it was less everything but it was the sort of thing you can't let slip away. When someone asks you what you do for a living, and you can say, 'I'm working as an air-filtration executive,' or you can say, 'I'm trying to find the biological basis of consciousness'—which would you choose?"

{ FIVE }

Stephen's Claw

After Christmas of 1997 Stephen was too busy to think about his tired hand. He would rest it when the house was done. Everyone in his family pitched in with woodworking, painting, gardening. In February of 1998 the weather turned apocalyptically bad because of El Niño, which is named for the Christ Child because it often appears around Christmas. It starts with a flow of strangely warm water off the coast of Peru; but everything in the weather system is connected, and the Child troubles the weather around the world. El Niño has come to the Pacific more and more often in recent years, a trend that may be driven by global warming, which is a problem that John Heywood was thinking about that year as he wrote his book about two-stroke engines. If John could improve the efficiency of the world's two-stroke engines so that they burned less oil and gas, he might help slow down global warming—and in that way he might even hope to do something about the weather.

Peggy clomped around in the mud, landscaping the front and the back yard with plants and vines. They had found some sod. The Mexican craftsmen her sons had hired sat on the roof in the rain and watched her. They laughed to see Peggy slogging through the garden with plastic trash bags wrapped around her legs up to the knees. They called down to her from the roof.

Excuse me! Don't you have an academic degree?

On February 19, the real estate section of the *Palo Alto Daily News* ran a feature about the new house for sale on Forest Avenue: "This week's *Daily News* Open House has been a labor of love for two brothers who have spent almost two years taking a Palo Alto house apart—salvaging windows and antique brass doorknobs—and creating something entirely new."

The reporter praised the custom details that Stephen and his family had added, from the fine terra-cotta fireplace to the outlets in the upstairs bedrooms, which were wired for cable and eight phone lines—vital details in Silicon Valley. The cottage was now about to be offered for $925,000.

The paper ran a photograph of Stephen and Peggy in the new kitchen, standing in front of the cherrywood cabinets with tiny built-in lights, designed by Peggy and built by Stephen. The caption reads "A Family Thing." Stephen has his arm around his mother and his right hand rests on her shoulder. Even in the photograph the hand looks strange. It is too thin, delicate, and spidery for the hand of a big man. The fingers curl inward. Everyone in the family had noticed the change in it by then, and the three brothers joked about it while they worked. They called it Stephen's claw. Whenever the trigger of his power drill felt stiff, or the nail gun kicked in his palm, Stephen yelled, "Damn this claw hand! Damn this claw hand!"

He stayed as good humored as ever, but all sorts of fine jobs, like turning the screws of C-clamps, were getting harder for him to manage. To open the front door now, he jammed the key between his right thumb and his palm and rotated his whole forearm in a semicircle, turning the key with his arm instead of his wrist and fingers. His handwriting was getting bad, but then, it had always been bad. These were all minor inconveniences. He and his brothers were still young and sunny enough to feel that aside from the usual Generation X angst, nothing horrible had ever happened to any of them. Stephen's right arm was still very strong. He could swing a hammer without a problem. Lifting the nail gun was nothing, even though he had trouble pulling the trigger. "We were working so hard that I just figured, well, geez, this could be anything," he told me later. "It could be exhaustion, it could be a pinched nerve, carpal tunnel—you know, who cares what it could be? It doesn't really matter."

By now, all of the Heywoods were possessed by the house. Stephen's baby brother Ben screamed into the phone at the roofer and threatened him with his nonexistent lawyer. (The roofer tiled the roof.) Their mother labored out in the yard from sunup to sundown, as if she were back on the farm. Stephen himself worked gracefully under pressure. He kept his perspective even in the last emergencies. When one of his apprentices laid a window on the ground, took a step forward, then a step back, and put his foot through the old rippled glass, Stephen barely flinched. He consoled his apprentice.

One afternoon that month, Stephen was walking across City Hall Plaza in Palo Alto, swinging his arms, when he happened to bang his right hand against a mailbox. He glanced down at his hand. Then he did a double take. He held up both palms and stared from one to the other. Something was wrong with the shape of his right palm. He stood in the sidewalk looking back and forth between his hands.

Holy shit.

Between the thumb and pointer of the human hand, there is a muscle that allows the two fingertips to meet. It is the most important muscle in the hand because it makes possible the quintessential human movement of the opposable thumb, drawing the tip of the thumb to oppose the tip of the pointer. According to evolutionary theory, the opposable thumb may be the piece of equipment that made the difference for *Homo habilis,* or Handyman. From the opposable thumb to caves and flint axes; and then to *Homo sapiens,* Man the Wise; then villages and statues, cities and bridges, roads and engines, vials and syringes, keys and locks.

Stephen held up his left hand with the fingers together, the thumb pressed up against the pointer. In a normal human hand, the muscle makes a bulge, a small, fleshy, wrinkled bulge, right in the crook at the base of the thumb. This bulge is known to anatomists as the thenar eminence. Stephen did not know what it was called, but standing there on the sidewalk he could see that his right hand had lost it. His right hand looked as if the crook between the thumb and the palm had been carved away with a jigsaw.

This is serious. It's not just exhaustion. Something is wrong here.

The weekend before the house went on the market, the Heywoods did a last tremendous burst of work to get ready for the big open house. You opened a house in Palo Alto the day before it went on the market, and it always sold in a day. On Forest Avenue, it was an Amish-style finishing. The whole family was racing together. The night before the opening, Stephen, Jamie, and their father put up the banister at the head of the stairs.

On February 28, 1998, the house went on the market, and it sold the same day for $975,000, fifty thousand over the asking price—almost a million dollars. Jamie thought it was probably the highest cost per square foot in Crescent Park. His brothers had paid $325,000 for the place, so they made a gross profit of roughly 300 percent.

The Heywoods celebrated together. Peggy asked Stephen if she could be his partner on his next house. If Stephen could find one to renovate in Boston, she would put up the money, and he could teach her to work with her hands.

As soon as the house was sold, Stephen went to see a doctor, although he still did not feel terribly worried. "Stephen is not one to go to the doctor. If he has a gash in his hand two inches deep, he'll sew it up himself. A doctor's the last thing on his mind," Ben explained to a television reporter long afterward, when the story of the Heywood brothers had become famous. "And Stephen was carrying the whole house around in his head. So he was mentally fatigued as well as physically."

The doctor sent Stephen to a hand surgeon, who sent him to a group of neurologists. The problem was simple, the atrophy of a muscle, but a thousand things can cause that. Sometimes, in rare cases, the cause is psychological, emotional, spiritual. The young Sigmund Freud wrote a paper on organic versus "hysterical" paralysis at the suggestion of Jean Martin Charcot, the great French neuropathologist who first described amyotrophic lateral sclerosis, ALS.

Cases of hysterical paralysis are curiosities in the medical literature, although the neurologist and writer Oliver Sacks once experienced the phenomenon himself. Sacks was hiking alone up a steep mountain in Norway when he rounded a boulder and came face-to-face with a bull in the path. Sacks panicked, ran back down the slope, fell, and hurt his leg badly, paralyzing it. A surgeon repaired the damaged muscle but after the operation Sacks discovered that his leg was not only still paralyzed, it no longer felt like his own, as if it were not even attached to him. "The muscle was toneless," Sacks writes in his memoir about the experience, *A Leg to Stand On,* "as if the flow of impulses in and out, such as normally and automatically maintain muscle tone, had been completely suspended. The neural traffic had stopped, so to speak, and the streets of the city were deserted and silent. Life—neural life—was suspended for the moment, if 'suspended' was not itself too optimistic a word." Sacks had to wait a long time before the signals got through again and he had two legs to stand on.

Stephen's doctors did nerve-conduction tests. They ruled out a pinched nerve, carpal tunnel syndrome, and many other possibilities, including psychological paralysis. They thought Stephen might have benign focal amyotrophy, also known as Hirayama syndrome, which is an inexplicable weakening of a single limb. The doctor who led the group told Stephen that his nerve damage would probably never heal. If he did have Hirayama syndrome, the weakness would not progress. His condition would have to be watched. If it did progress, that might mean that Stephen had the beginnings of ALS. The neurologist mentioned that possibility very casually, and Stephen played it down when he got home. He had Hirayama syndrome, and it was just a nuisance.

Oh, it's this weird thing that was discovered in Japan, and there you go. And it's no big deal, your arm gets weaker for a year and a half, and then it stops.

He was fine about it. He could learn to build houses with his left hand.

Of course, you always want two hands. But you don't need two strong hands, you need one strong hand. It's only when you have two weak hands that you're screwed.

While Stephen sold the house and talked with neurologists, in the spring of 1998, Jamie and Melinda moved to La Jolla, and Jamie started his new job at the Neurosciences Institute. Their apartment on Pearl Street had a sliver of Pacific view. They could sit in the evenings and watch the sun set through the waving fronds of palm trees. They could hear the waves all night.

Up on the mesa at the institute, Jamie was pleased to see that his new office had a beautiful redwood desk with green, architect-designed file drawers underneath it. The desktop computer was top of the line, the latest, fastest machine, with a big nineteen-inch monitor. Because Edelman envisioned the place as a scientific monastery, his architects had designed the scientists' offices to look like monks' cells. But they were stylish cells with poured concrete walls, and each cell had its own small mystic figure impressed in the concrete. Jamie had seen that kind of design element in his architecture magazines. The design in his own office wall was a series of dots, a sort of mystic ellipsis, significance to be determined.

Rather than hide Jamie away in a business office or in administration, Edelman had put him in the middle of the theory group. "He just dropped me right in like I was one of the scientists," Jamie says. "I think no one knew what to make of me—including myself, a little, at the beginning." He had never been a scientist, and he did not look like one. The neuroscientists' uniform at the institute was a polo shirt and a pair of khakis—they looked as if they had all crawled out of The Gap. Jamie wore expensive jackets and ties that made the scientists smile among themselves. *He dresses better than our director—and Edelman wears a suit every day.*

Scientists at the institute share their offices. That is another Edelman touch, meant to encourage conversation and cross-fertilization. Jamie's office mate was a molecular biologist named Joe Gally, who had been Edelman's first graduate student back at Rockefeller. Joe Gally helped do some of the early work on the immune system that led to Edelman's Nobel Prize. Then Joe went off to travel the world. When he returned, he got a job teaching biochemistry and cell biology at a medical school in Tennessee. At a certain point Joe lost his job in a departmental shuffle, and he called Gerald Edelman. *If I can't teach, at least I want to learn,* he said. Now he was deeply involved in the life and science of the institute. He was absentminded about time, space, clothes, almost everything but molecular biology. But he still missed teaching, and he was a mentor to many of the institute's younger scientists.

Joe was one of the legends of the institute. Directors of other research laboratories told stories about him. In the folklore, Joe was said to have the powers of memory of an idiot savant. Edelman was rumored to have hired him to read every page of every issue of virtually every biological journal as it was published online. Although that was not really true, Joe Gally did serve the institute's neuroscientists as a living encyclopedia. They called him the Walking Library.

The office that Jamie shared with Joe was right across from the espresso machine—the caffeinated nerve center of the institute, the best spot in the building for meeting neuroscientists. Jamie could meet the institute's scientists over coffee and then explore the institute by jumping from office to office and lab to lab. That was perfect for Jamie. He also went to the lectures of visiting scientists and the weekly meetings of the journal club, where neuroscientists at the institute took turns telling each other about new and exciting research papers.

Because Jamie had never taken a biology course, wandering into those talks was a little like walking into the courtyard of the Neurosciences Institute. He got his first confused, dazzled impressions of the spectacular invisible architecture that most people will never see. The human body is home to each one of us and it is also one of the most exotic places on the planet.

Jamie knew, of course, that living bodies are made of cells. Cells are to the human body what bricks or beams are to a building: the basic units of living architecture. Excluding viruses, single cells like an amoeba or a bacterium are the smallest living things on earth, and a human body is made of about 300 trillion cells living and working together. Ten thousand cells in a row would stretch several inches, about the length of the lifeline in a man's palm. They are called cells because one of the first scientists to see them through a microscope, Robert Hooke, in the seventeenth century, was fascinated by architecture. Hooke studied a thin slice of cork and saw innumerable tiny chambers, and he was reminded of monks' cells in a monastery.

Most of the lectures at the Neurosciences Institute were about the action one level down, inside cells, in the complicated Tinkertoy structures that we call molecules, of which the double helix is the most famous example. If cells are the building blocks of living bodies, then molecules are the building blocks of cells. Ten thousand sugar molecules in a row would stretch from one end of a cell to the other. And each molecule is made of atoms—tens, hundreds, thousands, sometimes tens of thousands of atoms.

Jamie was intrigued by the molecular gadgetry in the human body. In exploded, false-color views, floating in black space, these molecular machines look beautiful and strange. From an engineer's perspective, life looks like nothing but molecular gear. Some molecules inside a cell work like pumps, some like valves, some like filters, some like pliers and screwdrivers, some like clocks. Where anatomists a few hundred years ago argued about the mechanisms of our body's organs, now they argue mostly about the workings of our molecules. In muscle, for instance, there are molecular motors that drive the action, and these motors can be dissected and examined. Biologists argue about whether one motor pulls at a muscle fiber by hauling it hand over hand like a sailor; or whether the molecule pulls at the fiber by bowing down, pulling, stretching up, and then bowing down and pulling again, like an inchworm.

Jamie thought about Stephen. Somewhere in there, probably at the very smallest scale of the architecture, something had gone wrong. In the beginning maybe just one molecule had gone missing, and now Stephen was losing the use of his right hand.

In the evenings, Jamie browsed through one of his new boss's books, *Bright Air, Brilliant Fire: On the Matter of the Mind.* It was a new sacred book, not to be read in the bathtub. Jamie tried to visualize the gear inside a human brain at the level of cells and then at the level of molecules and atoms. For breaks, he played his favorite computer game, Quake, one of those combat games in which an action hero runs across the faces of burning planets, killing enemies and fighting to save the galaxy. Jamie, Stephen, and Ben were talking two or three times a day, as always. Sometimes they played Quake on the Web while they talked, arm wrestling across space, dueling a few hundred miles apart.

{ SIX }

Ponyenka

In February of 1998, while Stephen Heywood sold his house and went to see his first neurologist, and Jamie Heywood joined the Neurosciences Institute, my mother had her first appointment with a neurologist in Providence, Rhode Island. She was worried about her memory. The neurologist, Stephen Salloway, gave her a standard set of tests for cognition and memory problems. Like doctors she had seen before, Salloway could not make a diagnosis.

One week after that appointment, my mother had some trouble walking. She told my father that her legs hurt. Then she began to stumble and fall.

One fine Sunday morning that spring, I drove to New York City with my two boys to spend the day with my parents. My father and mother had driven down from Providence to spend the weekend in a hotel in the city. My wife, Deborah Heiligman, stayed home that day. She writes children's books, and she had a deadline.

I met my parents at the Metropolitan Museum of Art. We showed the boys the halls of medieval armor. It was one of the first brilliant days of that spring, and the sun hit our eyes when we came out and started down the long flight of stone steps toward Fifth Avenue. My mother's sister Doris and her husband Mort had met us at the museum, and they were walking down the steps on my father's side. I was on my mother's side, and Aaron and Benjamin were just in front of us, with Benjamin, who was nine, reaching backward a little awkwardly to hold his grandmother's hand. We had gotten most of the way down toward the sidewalk when Benjamin hesitated and half-turned to look back up at her.

My mother was seventy-four, but she looked much younger. She kept fit playing tennis and bowling, and she was wearing sneakers. Benjamin had not pulled at her hand, and she should have been fine. Still, she faltered. I took her arm just above the elbow with enough force to steady her. But something was wrong: She was collapsing toward the steps like a rag doll, or like a marionette whose strings have snapped. I managed to keep her from toppling forward, but as she sank down onto the steps I fell forward myself, thrown by the unexpected weight of her free fall. It was a good thing we were already most of the way down the steps. I somersaulted, head over heels, and hit the sidewalk spread-eagled on my back. The camera that was slung around my neck thwacked down an instant later, right next to my ear.

The crowd made a circle around me, but I jumped back to my feet. I was all right. Up on the steps, my mother and the rest of the family looked all right, too. My father was focused on my mother's fall. I could tell that neither of them had seen me go down. My mother was gently laughing off what had happened, to show my father that she was fine.

Even my old camera seemed fine. The barrel of the lens was dented, but the shutter still clicked.

Back in Providence, Salloway, the neurologist, did more tests, but he was not sure what was wrong with her. Many neurological diseases have distinct names, but they are really constellations of problems that shade into each other. That makes them hard to tell apart. It also suggests that if neurologists could understand what starts just one of the nerve-death diseases, they might understand many of them. Some neurologists hope or dream that there will turn out to be just one problem at the start, and when they find it, they can zap it.

One sign that many of these diseases are linked is the strange case of the Chomorros, the Polynesians on the South Pacific island of Guam. In most years in the second half of the twentieth century, one out of three on Guam died of nerve-death disease. Sometimes the disease looked like ALS, sometimes like Parkinson's, in other cases like Alzheimer's or a rarer condition, progressive supranuclear palsy, PSP, also known as Steele-Richardson-Olszewski syndrome, after the three doctors who first identified it in 1964. The youngest of those three doctors, John Steele, moved to Guam early in his career and has spent decades there, convinced that the Chomorros were the missing key to them all.

Steele knew that whatever caused that epidemic, which was known locally as lytico-bodig, it was almost over. The only Chomorros who died of it had been born on the island before or during World War II. Baby boomers and their children and grandchildren did not get lytico-bodig. So if he hoped to solve the mystery, time was running out on Guam.

Early in the 1990s, Oliver Sacks visited Steele on Guam and joined him on his rounds, seeing patient after patient who had what looked to Sacks like ALS or PSP. Even for the professional these are not easy encounters. Sacks writes: "I felt drained by seeing these patients with lytico and bodig in their final, terrible stages, and I wanted desperately to get away, to lie down and collapse on my bed, or swim again in a pristine reef. I am not sure why I was so overwhelmed; much of my practice in New York involves working amid the incurable and disabled, but ALS is rare—I may see only one case every two or three years." Steele, who kept up a jovial booming manner with his patients, wept in private because after thirty years he still could not do anything to help.

Guam is not the only case like that. There is another on an island off the coast of Japan; and in the French West Indies in the late '90s another epidemic of nerve-death disease was discovered on the island of Guadeloupe. There the disease began the way it did on Guam, with clumsiness, unsteadiness, postural instability with early falls. The syndrome was ten times more prevalent on Guadeloupe than in Europe or North America, and it was confusingly variable. In one study, a team of neurologists looked at a random sample of 220 consecutive patients who had been seen by the neurology service at Guadeloupe

University Hospital in Pointe-à-Pitre and diagnosed with what was being called atypical Parkinson syndrome. Of those 220 patients, more than 90 had some kind of parkinsonism, 58 seemed likely to have PSP, 50 had well-defined Parkinson's disease, 15 had ALS with parkinsonism, and one had probable multiple system atrophy.

So a large number of nerve-death diseases may be linked, and no one knows how many others are also linked but still unidentified. John Donne wrote on his sickbed: "O miserable abundance, O beggarly riches! how much doe we lacke of having *remedies* for everie disease, when as yet we have not *names* for them?" Because many neurological diseases shade into each other, many of the names we do have are probably spurious; and the first beginnings of neurodegeneration are usually impossible to trace, as James Parkinson observed in London in 1817 in his classic "Essay on the Shaking Palsy": "So slight and nearly imperceptible are the first inroads of this malady, and so extremely slow is its progress, that it rarely happens, that the patient can form any recollection of the precise period of its commencement."

My mother did think she knew how it started. She and Dad spent summers in a mountain cabin near Marlboro, Vermont, for the music festival. On a hike up there in the late '80s, she had a bad fall. She slipped on wet leaves and pine needles and went sliding down the path fast in a sitting position until she hit a tree. Her legs flew around it to either side and her forehead whacked into the trunk. When she felt her memory begin to fail a few years later, and her thoughts began to get dark and painful to her, my mother blamed that fall. People with neurodegenerative diseases often report that they had had an accident or a period of great stress about two years before symptoms appeared. So maybe that fall did the damage, but no one really knows. Maybe it was the onset of the disease that made her fall.

She got moody, absentminded, a bit paranoid. She thought she had Alzheimer's. Once when I was visiting my parents in Providence, and my mother and I were alone in the kitchen, she pointed at the kettle on the stove. It was steaming. "Did you put that kettle on?" she asked. She was looking at me sidelong, strangely, as if the answer to her question would be cosmic and she almost could not bear to hear it.

No, but so what, Ma? You're fine.

When she was young, her nickname was Ponnie, or Ponyenka, which is a Polish endearment that means Little Miss. Her father was a tailor who had a shop on Manhattan's Lower East Side. His greatest success was the Shirley Temple dress, which appeared on the cover of *Life* magazine in the middle of the Depression. So my mother and her sisters dressed in style in Bensonhurst. Their family name was Mensch, which means, roughly, a good person, a human being in the best sense, a deep, broad character, as in the Yiddish proverb, "Ten lands are more easily known than one mensch." My mother's given name was Florence, after the loveliest city in the world. So, a lovely human being. But there was also something a little fragile and nervous about her, even when she was young, and everyone who loved her felt protective of their Ponyenka.

She married a graduate student at Columbia: an engineer, Jerome Harris Weiner, my father. He became a professor there, and taught engineering and applied mathematics. He did well in the boom after the war, when Vannevar Bush called science the Endless Frontier. It was almost as good a time to be an engineer then as it was at the end of the century. My father took our family on sabbatical years abroad in Italy and Israel, with weeklong Atlantic crossings on the *Queen Elizabeth* and the *Independence.*

When he moved from Columbia to Brown University, in Providence, my parents bought a Victorian on College Hill, a few minutes' walk from the university. The house was three stories high and formal, with two staircases and a butler's pantry. A speaking tube ran from the kitchen to the old servants' quarters on the third floor, which became my aerie as a teenager. My mother planned and oversaw the restoration. When she was done, I think Stephen Heywood would have liked the house. It had a fine front staircase, a wall of books in the living room, a dining room with a long sweep of polished walnut floorboards. She decorated the house with antiques and also with a few of the heavy, dark, Germanic pieces from Bensonhurst, because at that time my mother was the only one in her family who had room for them.

Providence is four or five hours' drive from New York City. When my aunts and uncles came to visit, they thought the house was too far away. They may also have thought it was a little too much. My parents always seemed so *perfect,* cousins on both sides told me after my mother got sick, and there was something about that line I hated. I know they loved my parents, and I am sure they were worried about me, but what I heard was, "Lo, how the mighty have fallen!" Forty miles away, in Newton, Massachusetts, the Heywoods would soon be receiving the same line of sympathy. One of their cousins told Melinda that the Heywood family had always seemed almost too golden, too blessed. Melinda brooded. It was as if their cousin was saying, "You finally got yours."

Lucretius put that problem on papyrus, along with so much else. "Pleasant it is, when over a great sea the winds trouble the waters, to gaze from shore upon another's tribulation: not because any man's troubles are a delectable joy, but because to perceive from what ills you are free yourself is pleasant."

Well, that is a human resentment on both sides, on the boat and on the shore: one of those eternal human feelings that make us all eternally uncomfortable, even when we believe that we are the one who is up there on the hill.

My parents certainly had gotten way out of Brooklyn, in the big house on College Hill, and they paid for their isolation once my mother got sick. Late one night my mother called one of her sisters back in New York. She said hello and then started to cry. She did not say a word. She would not stop crying.

Ponnie, what is it? What is it, Ponnie?

By the late 1990s, I was driving up Route 95 to New England as often as I could, often with Deb and the boys. At home in

Pennsylvania I spent hours on the phone with my father discussing doctors and medicines. Sometimes I began my days at dawn writing down early family memories. I kept going back to our year in Italy, when I was five, and our year in Israel, when I was twelve. We used to have family picnics at the Caves of Carmel, where my kid brother, Eric, and I found flint tools made by Cro-Magnons and Neanderthals, and where we placed our palms against palms that children had painted on the walls of the cave tens of thousands of years before we were born.

In neurodegenerative diseases, nerves die. The differences in symptoms between the multitude of related diseases depend in large part on which nerves die first and which ones die next. As her illness progressed, my mother became very difficult, and she drove my brother away. But I was the first son, and I could still revive all those ancient memories. She was a children's librarian and when we were growing up she always had the right book. I could hear her voice singing to us: *Toorah loorah loorah.* My memories went all the way back to the crib. In one of them I am looking through the crib bars, or sometimes through the five fingers of one hand. I move my head a little to the left, then a little to the right, and watch a light across the room appear and disappear. It is more than just something to do. It is a game, it is almost a play, a story about light and dark, here and gone. New pairs of eyes must have played that game in every generation and made some kind of parable out of it. Martin Buber heard a saying from a disciple of Rabbi Nachman of Bratslav, who had passed it down from his great-grandfather, the Baal Shem Tov: "Alas! the world is full of enormous lights and mysteries, and we shut them from ourselves with one small hand!"

Something was very wrong with my mother, although she still kept herself fit, watched what she ate, kept up her bridge and tennis, and dressed with style. She was still Ponnie, Florence, the pretty one, her father's daughter, and she looked younger than her age. But she lived in dread that her disease was about to tear her apart.

Late one evening I visited my parents in the tall old house in

Providence. Was this during the same season when she stumbled on the steps of the Met, or was it earlier, or later on? Most of our family disaster is a blur to me now. In houses, even big houses, the most important conversations sometimes happen in the most cramped, out-of-the-way places, as if there anything goes. The conversations that are not meant to take place somehow find their way out in corners or in back stairways under sixty-watt bulbs, where they are shared as if by conspirators. My mother and I were standing at the foot of the back stairs by the coatrack. In the stark light and dark of the landing every object looked slightly surreal, the tennis rackets and tubes of tennis balls, the rubber boots way in the back, the filing cabinet for her favorite catalogues. We had turned out all the other lights on the ground floor, getting ready to go up, when I stopped and turned. I told her that I could see that she was suffering, and I said, "Sometimes you must think about ending it."

Her expression when she looked at me was startled, abashed, and frank at the same time. Her face was confessing that she did think about suicide all the time, and that she put the thought away again and again. She would never kill herself, she told me. "I know what kind of legacy that would leave you and Eric."

In the dim light, I asked her to promise me that if she ever began thinking seriously of killing herself, she would call me first. I thought there had to be something that medicine could do for her, and I wanted her to hold on.

I will never know what she thought of our talk that night, or if she even remembered it. She was so shut down in her own weather by then that she was not looking to me or to anyone else for much. My father was still real to her; the rest of us were already far away. I was stepping forward, pushing through that distance, getting in her face. It was the boldest I had been with my mother since I left home. I was offering a contract. *You hold on. I will get help. Stay with us.*

But we did not even have a diagnosis. If I had known what she was in for, I would not have begged her to hold on.

{ SEVEN }

A Pang of Fear

Jamie was intrigued by biology, but in his first few months at the Neurosciences Institute he felt very stretched. He learned a little genetics, cell biology, molecular biology. Complexity theory. Neural nets. He was asked to see if some of the software the institute's cognitive neuroscientists were developing might interest toy manufacturers: Maybe it could help them develop new and improved Barbie dolls and G.I. Joes, toys with artificial intelligence. "But their stuff was so much more sophisticated and elaborate than you could handle in a toy," he says. "NOMAD runs on sixteen computers and it can barely recognize a block. It's an amazing achievement, but. . . ."

Jamie had a few private ideas, too. He had million-dollar dot-com schemes and business plans that he kept in the drawers of his redwood desk, including a concept for getting commercials onto computer screens. Some of his best ideas came in the evenings while he played computer games with Stephen. When they played and talked on the phone, they could see each other's avatars dashing through the combat zones on the screen. Jamie thought that was amazing. There he was in La Jolla warning his brother in Palo Alto, *Watch out for that guy behind you!* At the institute, he met with a computer scientist from MIT who had designed software for reading and interpreting human facial expressions. Jamie wondered if he could write a module that used webcams to convey people's expressions onto their avatars. Then when he played Quake, he would see Stephen's reactions on the screen and Stephen would see his.

Ideas like that could make the right entrepreneur 10 million dollars overnight in 1998, and Jamie loved to spin get-rich-quick schemes with his new friends at the institute. He loved to talk, and he had always learned best by talking, not by reading books. Stephen, who had turned away from books, could read a page in a flash and remember everything he read. But Jamie, who revered books, was dyslexic. He often scribbled down a phone number with two digits reversed and then reversed them back when he made the call. Like many dyslexics, Jamie was also dreamy, hyper, and distractible. In classrooms he used to wander from desk to desk and chat, which drove his teachers crazy. He did not do well enough at Newton North to get into MIT, and he flunked out of his first year at Carnegie Mellon.

"I came home," Jamie says. "Figured I'd work for a year, get my head on straight. Figured I'd get a job selling computer systems, and retake some classes."

"What he *said* was, 'I think I'll transfer to MIT!'" says John Heywood.

John was sure that Jamie was dreaming another one of his dreams. He assumed that MIT was out of reach for a student with Jamie's record. At least, that is how father and son tell the story. John had a friend in admissions, Dan Langdale, and he asked Dan to see Jamie, thinking that his friend would tell his son to wake up.

"My father seized the opportunity to teach me that there is no recovery from terrible errors," Jamie says. "Dad had every expectation that Dan Langdale would tell me I was going down hard."

"Yes. No one will ever believe that I didn't expect that Dan wouldn't gently say—"

"But Dan said, 'Of course.'"

Langdale gave Jamie a chance. He let him take two MIT courses as a special student. When Jamie did well. Langdale let him in.

"I probably hold the record for the lowest average of anyone to transfer to MIT," Jamie says. "The funny thing was, my father had wanted me to learn a lesson. Instead I learned that rules can be bent!"

Jamie was bending the rules again by quitting engineering to work at a biological think tank; and he was bending the rules there, in a way, by trying to get the gist of biology, skimming from subject to subject. That was his job—but it is not how scientists work. In his monk's cell, while his office mate Joe Gally did abstruse theoretical work, or scanned the Web and absorbed the day's discoveries in molecular biology, Jamie studied the research papers of the neuroscientists around him, with help from Gally and a growing shelf of books, including Edelman's *Bright Air, Brilliant Fire* and a handy guide called *The Human Brain Coloring Book.* He tried to talk the talk at his meetings with Edelman. On his fast computer with the big screen, he ran the scientists' models of the workings of the brain—highly mathematical, highly theoretical—looking for commercial applications. Maybe one of their computer models of consciousness could help predict the rise and fall of the stock market. Maybe he could help them turn that idea into a company. Now and then, as a favor, he designed his new friends a piece of lab apparatus, or coached them in the principles of business plans, dot-com entrepreneurship and salesmanship. *It's got to have WIIF'M,* Jamie said. That was one of his favorite acronyms: WIIF'M, What's In It For Me. Show the other guy what's in it for him and he will buy your idea every time. Almost every day he stopped by Ralph Greenspan's office. *I'm going to make you a rich man!* Jamie liked to say. *I'm going to make you all rich!*

All three of the Heywood boys had a kind of something-for-nothing charm in those days. They were at home in the new California Gold Rush. When Jamie visited Stephen and Ben in Palo Alto, they hung out at a pool bar and restaurant called the Blue Chalk, one of the hottest and toniest spots in Silicon Valley. The three of them knew everyone at the Blue Chalk. Jamie once boasted to me that his brothers could walk in on Saturday night when there was a three-hour wait for a table and have one in fifteen minutes. "And I don't think that

Ben ever bought a drink there—he got free beer from all of the bartenders. He tipped extraordinarily well, which helped."

Their cousin David Searls ate there with them one night when he was in town. "They've always known how to have a good time," he says. "Stephen was especially well liked by the female servers. God, yes. He had no trouble attracting women."

Searls is fifteen years older than Jamie. He has degrees in philosophy, life sciences, developmental biology, and computer science. Back then he was on the faculty of the University of Pennsylvania, working on genomics. Jamie used to call him for advice as he tried to find his way into his new job. Searls has known all three of the Heywood boys since the week they were born. He has watched them grow up on the farm in South Dakota and on the beach at Duck, and he has never quite understood them. "The one word is 'irrepressible,'" he says. "My uncle John is the quintessential academician, not to mention being British, so it always struck me that he must have always wondered," Searls sighs, "about marrying and mixing his seed with the American strain, since this is what resulted. I don't know how he feels, because he's pretty quiet about his feelings, but I've always wondered if he didn't quite know what to make of having all these wild children."

Like Searls, most of the Heywoods' cousins have conventional careers as academics and other white-collar professionals. But Stephen was happy as a part-time carpenter, Ben had quit engineering to shoot for Hollywood, and now Jamie had quit to dabble at the edges of biomedicine. "It's courageous to do what makes you happy and follow your muse and that sort of thing," says Searls. "But all three of the Heywood boys followed their muses to the exclusion of any concern about tradition or the rigorous approach to life that us Midwesterners are supposed to have. Some of us have to work for a living."

Searls says he had warned Jamie that without formal training, much less the certification of a scientific degree, he might find it hard to get by at the Neurosciences Institute. "His intuitive style had clearly taken him a long way," Searls says, "but I felt at the time, being trained as a scientist myself, that it would only go so far. There is a place for rigor and conservatism."

And Jamie did worry. From early on, he worried that Edelman was disappointed with his progress. Although neither Jamie nor Gerald Edelman will say much about it now, some scientists at the institute wonder if there was a clash of styles. Jamie has a genius for picking up the lingo of a place and fitting in—he is a chameleon. He also likes to dominate a conversation. From the beginning, he may have tried to talk as if he *knew.* Old-school scientists like Edelman have almost a physical revulsion against claiming to know more than they know or claiming to have done more than they have done.

On the other hand, when Jamie explained business start-ups, Edelman may have tried to talk as if *he* knew.

Jamie agonized that spring, but at the same time he felt sure that he would triumph in biology in one way or another. He could make anything work. He was a problem-solver. He was a Heywood. All this was part of Jamie: pride, confidence, hubris, convictions of grandeur, dreaminess, family loyalty, compulsive salesmanship. When Jamie flew, he liked to talk a stewardess into bumping him up to first class. Once he badgered me into trying it, too, but I did not get far.

"This is a coach-class ticket," the stewardess said.

"You're right."

I told that story to Searls. "Yeah," he said sourly. "*I* have to fly a couple of hundred thousand miles before *I* can do that."

Jamie could doubt himself. But he also had faith in an invulnerably charismatic charm, a charm that could get him into or out of anything he wanted.

All through the spring of 1998, while Jamie chatted with biologists who study the human body's nervous architecture, Stephen tried just as hard to put the subject out of his mind, even though his right hand reminded him every other minute that something was wrong. His buyers on Forest Avenue loved him, and even after the sale they paid him to keep working on the place. They wanted a doghouse in the backyard in the same custom-color stucco as the house. That spring, Stephen built them a twenty-thousand-dollar doghouse. This took a lot of stonework, and he began to notice weakness in his right hand as a whole. Now it was not just his thumb and forefinger. He had problems doing all sorts of little jobs, like wringing out a sponge. When he put his right hand under cold running water it would freeze up and he would have to straighten out the fingers one by one with his left hand. To hold a wrench or a drill, if his right hand was chilly, he had to wrap it into place one finger at a time with his left hand. When he wanted to put a tool down again, his fingers would be stuck around it, and again he would have to pry them straight one by one. Still, his right arm was as strong as ever. He could set a big piece of wood resting in his palm, and, using the hand like a hook, carry the wood up a ladder with no problem.

When Jamie flew in, Stephen enjoyed having him around, although he did not care to hear much about science. Later on, Stephen would talk about that spring in his usual I'm-not-an-intellectual style. "The guy who bought the house, he works for a biotech. So he and Jamie had lots of conversations talking gibberish."

Stephen felt he had earned the right to slack off again. "It's not like we made gazillions," he would say, "but we made enough to relax for six months." So that summer he and Ben celebrated the house sale with a trip to Europe. They stayed in hotels instead of youth hostels, and they tooled around Monte Carlo, Stockholm, St. Petersburg, Prague, Budapest, Vienna, and London. In Monte Carlo, they dressed their best and played blackjack, the only game in a casino where the odds against the gambler are close to even, if the gambler plays a perfect game. Stephen was not a bad gambler. The two brothers won a thousand dollars over two nights.

With his left hand, Stephen wrote a cheerful postcard to one of his neurologists in California. "P.S.," he scrawled. "How do you like my left-hand writing?" As a gag, he spelled every word wrong. He hoped the doctor could tell he was kidding.

Along the way, he and Ben met up with the rest of their family for a holiday. House building had shown the Heywoods that they enjoyed each other's company as much now as they had when the brothers were boys. John and Peggy had rented an old Tuscan farmhouse just south of Siena, near the village of Montestigliano. There John worked hard on his book *The Two-Stroke Cycle Engine.* He also consulted for Ferrari for a few days, so there was an Alfa-Romeo to drive on the roads of white gravel that wind through the hills and the olive groves between Montestigliano and Siena.

At the old farmhouse, the brothers and Jamie's wife Melinda lounged by the pool. Jamie arm wrestled with Stephen and beat him again with his right hand. They ate wild boar stew. On a day trip in the old region of Cinque Terre, the brothers posed with their parents for a family snapshot, the five Heywoods standing tall and grinning in dark sunglasses in front of five stone towers. They went rambling in the woods around the farmhouse and saw places where boars had rooted for truffles. In Siena a boar's head hung outside a butcher shop. It did not look fierce in the sunlight. Someone had placed a pair of wire-rimmed spectacles on its snout.

After dinners at the farmhouse, Stephen and Jamie went out in the woods to hunt for a wild boar, a *cinghiale.* Melinda wrote up their adventure in her journal. Jamie took a walking stick, Stephen took a flashlight, and they marched into the dark hoping to get the wits scared out of them. If they did meet a boar, Stephen planned to climb a tree. Jamie counted on beating it off with his stick.

They were both a little nuts, Melinda wrote. She made longhand notes for an essay that she would later publish on the Web, "In Search of *Cinghiale,*" by Melinda Marsh Heywood. Besides her careers as a belly dancer, a circus performer, and a scholar of medieval French literature, she planned to be a writer. She wrote: "They are the kind of men who remain firmly in place with mad gleamy smiles when they hear that the eye of a hurricane might soon be passing directly over their vacation home and the rest of the island is evacuating." They had done that at Duck a few years before.

Whatever is dark, whatever is unseen, whatever is unknown, the boys went there through action. They were not writers like Melinda, not introspectors. They went to the edge through games and action.

Jamie and Stephen did not meet a wild boar that night, and the next night they tried again. Out in the woods, Jamie made Stephen turn the flashlight off to help their eyes adjust to the dark and to keep the quest as scary as possible. Stephen told Melinda about that to make her laugh. She jotted down a quote from William James: "In civilized life . . . it has at last become possible for large numbers of people to pass from the cradle to the grave without ever having had a pang of genuine fear."

On their last night out alone in the woods, just as they were about to quit and turn around, the brothers heard a snuffling sound in the dark. They stood still and held their breaths—but nothing happened. They trudged back toward the yellow lights of the Tuscan farmhouse without having met the *cinghiale.*

Melinda wrote it all up in mock-epic style. Someday, the brothers would have to come back. Someday they would meet that monster in the dark. "For now, however, the boar dons spectacles, cracks open the *Inferno* with a cloven hoof, and waits for the dauntless duo to return."

{ EIGHT }

Back to the Cave

In September of 1998, Stephen Heywood moved back into his old basement room in his parents' house in Newtonville, which he called the Cave. The Cave had been his first construction experience. With his mother's help, he had put in a carpet and a waterbed and slathered plaster on the ceiling.

Working mostly with his left hand, he began taking up the floor in his parents' bathroom and reinforcing the joists. To thank them for their help with his house in Palo Alto, he was building them a new master bath and bedroom. ("That was his idea, not ours," says Peggy.) Stephen still felt healthy. He was thinking of training for the Boston Marathon. He could hurry up the basement stairs on Mill Street pulling a shirt over his head without giving it a thought. He could broad-jump down the front stairs and land as lightly on his feet as he had when he was trying to amaze Jamie as a teenager.

Now that he had discovered what he wanted to do with his life, Stephen had also fallen in love. He and Wendy Stacy had known each other a long time—she had once gone out with one of his oldest friends, Robert Bonazoli. Stephen and Wendy had gotten together that summer, and the same night he pulled up with a U-Haul outside his parents' home on Mill Street, they had dinner in Cambridge. It was September 7, 1998, their first real date.

Wendy worked as an administrator in a few biological laboratories at Harvard. She was bright, cheerful, energetic, and optimistic, all good traits for getting along with Heywoods. There was also the attraction of opposites, because Wendy was as short and fair as Stephen was tall and dark. Her word for him was "hunky." They spent a lot of time together at her place, and Stephen felt very happy with her down in his Cave. He had hung up a big framed nautical map, "Duck and Beyond," on the wall, and some of his old oil paintings from college. "At Colgate," Stephen told me once, "everyone else was painting what they were painting. I was painting tools." Now they decorated the Cave: big hyper-realistic canvases of pliers, a plane, and a Swiss Army knife.

Stephen found the house on Mill Street a nice place to come home to, with his mother's flowers on the tables, his father's watercolors on the walls, and wavy old glass in the windows. He liked giving tours of the house with his mother. Every new guest of the Heywoods begins with a tour, and Stephen gave the carpenter's version. "The house was built in 1917. Typical colonial from the outside. These moldings are gumwood, which you can't get anymore. . . ."

Upstairs, he explained how he had taken the floor up in the master bathroom and reinforced the joists. "Too much support in retrospect," he would say.

"You can't have too much support," said Peggy.

Stephen's father was working sixty-hour weeks at the Sloan Automotive Laboratory, as always. But on weekend afternoons John worked side by side on the carpentry with Stephen.

Taking his time, Stephen hunted for a house in Back Bay, South End, Charlestown, South Boston. His mother still wanted to work on the next one as his partner. He figured that they could afford something at six hundred thousand, which was a stretch. He did not think Peggy was quite ready for that number, but he planned to prepare her gradually.

The master bathroom was taking much more time than Stephen had expected—in the end it took nine months to finish. When he told me that part of his story, later on, he seemed embarrassed. "At first I was trying to work by myself, which I should know better than to try and do, because I can't do it. I can't motivate myself to work without someone else there. Very hard to get started. Whenever I work by myself it's a dismal failure." Stephen would futz upstairs for a few hours and then drift down to the den to watch daytime TV, or down to the Cave to play Quake. He could still hammer with his right hand, or use a screw gun, even a nail gun, though it seemed to kick worse all the time. But his right arm was getting weaker and he began to see muscle weakness and thinness in his arm, all the way up to his elbow. Even his bicep and tricep looked thinner.

To follow up on his condition, Stephen made an appointment with Robert Brown, a neurologist at Harvard Medical School and Massachusetts General Hospital. Brown is one of the best neurologists in the country. He also happens to be the world's leading specialist in ALS, although that is not why Stephen chose him. His practice is busy, and Stephen had to wait a few months for the appointment. He still was not worried—he and Wendy were too busy falling in love. They had many dinners at Jae's, the little sushi place where they had their first date. They spent every weekend at her place, and many nights down in the Cave, and a few at little New England bed-and-breakfast places. On November 28, 1998, Wendy wrote in her journal:

> The year is so close to over. This year went faster than the last, but what a wonderful year it has been. . . .
>
> I have relaxed so much around Stephen. We are exclusive now and it is shaky and wonderful as we inch out each feeling and look for the other, touching cautiously, but happily. It feels right and it never has before with anyone. How happy I am with this man.

On December 16, 1998, Stephen reported to the clinical neurology lab at Massachusetts General Hospital for what he expected to be a forty-five-minute checkup—a few routine tests before his appointment with Doctor Brown in January. Almost one year had passed since he stood outside his first house with the key in the door.

At the lab, a young neurologist inserted a needle in Stephen's bicep and asked him to flex it while he watched the electrical output on an oscilloscope screen. The neurologist kept the screen tilted at an angle so that he could see it but Stephen could not. He tested both of Stephen's arms thoroughly with two-inch needles, then his legs, his groin, and even the back of his tongue. He also stuck a needle into the muscle between his right thumb and forefinger, where the palm had atrophied. He tried that spot again and again but he could not get any reading there at all. With an electrified skullcap, he jolted Stephen's body in a crescendoing series of shocks. All of these tests were extremely painful, and they took five hours. The neurologist kept calling in colleagues to look at the results.

Stephen tried to keep his voice casual.

Did you find any damage?

Yeah, I see some weird stuff here, the neurologist said. From the man's expression, Stephen could tell that he did not want to say another word. He told Stephen that Doctor Brown would deliver the diagnosis at his next appointment.

Stephen got himself out of the hospital and headed west toward Newton. On Storrow Drive, he picked up a cell phone that he had borrowed from his parents and called Jamie.

Jamie was working when the phone rang. His office mate Joe Gally had already gone for the day. He was alone.

Things had not gone as well at the Neurosciences Institute as Jamie had hoped. He had decided that the neuroscientists were too far ahead of their time for Wall Street. He still thought the best moneymakers he had going were the projects he kept in his desk drawers, the Internet start-ups. He could leap to glory someplace else. Melinda had not found a teaching job with her new PhD and she had never felt at home in La Jolla. The night before Stephen's call, she had been feeling as aimless and restless as Jamie. She had been waiting for a call for a job interview at the next annual meeting of the Modern Language Association, and she had gotten no calls. Her last entry in her journal the night before had dwindled down to a to-do list. "I should take Jamie to work tomorrow & get his presents." (Jamie was about to turn thirty-two.) "Return library books to UCSD . . . Buy wrapping paper. Keep pretending to try to write academic article so that I get other stuff done. I 'spose."

That afternoon, a venture capitalist in New York had phoned Jamie about a job. It was a much richer offer than the institute could ever hope to match. The call had made Jamie feel lonely and jangled. Usually he spent a good part of the day hanging around the espresso machine. Now he could not talk about what was on his mind with anyone out there.

Stephen was another reason that Jamie wanted a change of scene. At least carpentry gave his brother some distraction. For Jamie, working at the Neurosciences Institute meant thinking nine to five about all the things that can go wrong with nerves and brains.

He felt particularly fretful that afternoon, wondering about Stephen's checkup. While he waited for his brother's call, he worked on his résumé. Under "Experience," he typed on the first line, "The Neurosciences Institute," and added, grandly, in italics, *"Unraveling the Mysteries of the Brain."* Under "Interests," he typed "travel," "sailing," "construction," "renovation," and "family." Under "Objective," he put a question mark.

When the call came, Jamie grabbed the phone.

So what did they find?

Well, it doesn't look good. There's damage in all four limbs.

Jamie and Stephen tried to sound unflappable. They were The Boys: They kept their voices clipped and terse to spare each other's feelings. But they both knew what the news meant. The moment Stephen hung up, Jamie turned back to his desk. He closed his résumé. He started a search on his computer.

Then he typed the keywords "prognosis" and "ALS."

Of all the pathologies in medical textbooks, amyotrophic lateral sclerosis is one of the most painful diagnoses for a doctor to deliver and a patient to receive. It has been a death sentence ever since its discovery a century and a half ago by Jean Martin Charcot. In France it is still called Charcot's disease. In America it goes by the name of its most famous victim, Lou Gehrig.

In ALS, nerves in the spine begin to die one by one. These are what are known as the motor nerves, because they carry signals from the brain to the muscles of the body. Without signals from those motor nerves, the muscles wither and atrophy. As more nerves die in the spine, the body becomes progressively paralyzed. The damage travels up and down through the spinal cord until it reaches the brain stem, which controls the muscles that allow a human body to breathe and swallow. Even then, patients remain wide-awake and fully conscious. They can still see and hear, feel and reason. They watch their muscles wither away until they can no longer breathe, and they suffocate to death. "There does not exist, as far as I am aware, a single example of a case where, the group of symptoms just described having existed, recovery followed," Charcot wrote.

The search engine PubMed took only a second to scan its tens of thousands of biomedical references. Jamie's computer screen filled with the titles of scientific papers, and Jamie clicked on the first one. Now the screen displayed the abstract of a study from the University of Amsterdam in the Netherlands. He skimmed it. The study reported that most ALS patients die within two years of their date of diagnosis. Those patients whose first sign of weakness is a difficulty in breathing or speaking have the poorest prognosis, because in those cases the nerve damage has begun in the brain stem. Messages are no longer reaching the muscles in their chests and throats. They often die within months. Those patients with the best prognoses are young men with initial weakness in one limb and a long delay between the first sign of trouble and the diagnosis. Those patients often live five or six years.

With a shock, Jamie realized that he himself had detected the onset of the disease when he arm wrestled with Stephen at Duck. That was in July of 1997. He knew that no one else on earth could have been that accurate in measuring the strength of his brother's right arm. No one else on earth would ever have detected a problem then. The earliest that any doctor would probably have noticed anything would have been later, maybe that fall.

Seventeen months had passed since he had beaten Stephen with his right hand at Duck. So in the categories of the Amsterdam study, Stephen was a young man with initial weakness in one limb and a long delay between the first sign of trouble and the official diagnosis. That meant he might still have four years to live. But how long before he was paralyzed?

Sitting alone in his office, Jamie scrolled through abstract after abstract on the Web, trying to guess how long his brother still had to walk and talk, and searching for hopeful notes. He read about riluzole, the only federally approved ALS drug. Riluzole was reported to postpone death an average of three months.

It was a bad moment to be searching the Web about ALS. The Web was full of news about it in late 1998. Just that fall, Jack Kevorkian, "Doctor Death," had killed a man in Michigan on *60 Minutes.* Tom Youk, who restored and raced vintage cars, could hardly talk or write but he had convinced his family and Doctor Death and the television audience that he no longer wanted to live. Kevorkian had chosen to televise that killing because he was convinced that this was one case in which the world would have to agree with him about the merits of euthanasia. Youk was paralyzed and terrified as he watched the advance of his ALS. Kevorkian was now in jail for murder.

That year, about twenty-five thousand people had ALS in the United States. Five thousand new cases were diagnosed, and five thousand died. The number of people with the disease would be much larger if it gave its victims longer to live. On his computer screen, Jamie found ALS patients' Web pages recommending the removal of fillings from teeth. They posted a whole carnival of notes on the Web about healing circles, acupuncture, snake venom, and bee pollen. Some of these patients were already gone but their Web pages were still there.

One of the biggest bestsellers of 1998 was *Tuesdays with Morrie* by the sports writer Mitch Albom, the story of the last days of a Brandeis professor who died of ALS. The book had come out the year before—the same year that Stephen noticed his first symptoms. It was an inspirational book but it was also full of horrible news about what was in store for Stephen. "ALS is like a lit candle: It melts your nerves and leaves your body a pile of wax. . . ."

Morrie had lived and died in West Newton, Massachusetts.

As Jamie searched the Web, small thoughts kept threatening to swamp him. He had never really cared about arm wrestling. He did not give a damn about arm wrestling. But for years he had been trying to get better than Stephen at basketball. Now he would never have a chance to win one-on-one in a fair fight.

Not everyone could have done a systematic Web search like that just after hanging up the phone. Most people would have lost it. Jamie called Ben in L.A., and Ben curled up on the floor and cried. Jamie called Melinda, and she drifted out of their apartment and into a New Age bookstore, the Psychic Eye. Later on she wrote an essay about that moment. "It was the closest thing to a church on Pearl Street," she says. She wandered past sticks of musky incense and mystic candles, shelves of paperbacks on feng shui and homeopathy. She remembers skimming a book called *Prayers for Every Occasion,* but she could not find a prayer for this one.

I remember how I felt not long afterward when Stephen Salloway, my mother's neurologist in Providence, finally made a diagnosis of her condition. Salloway said it might be Lewy body dementia, which is one of his special research interests. LBD is another rare, progressive neurodegenerative disease, also fatal and incurable. It usually kills after about seven years. My mother had been sick for at least a year by then—maybe even for nine, if you counted the way her mood had darkened after she fell in Vermont. Now a doctor was telling us that he could do nothing to save her. Having a diagnosis to give out to my mother's sisters and to the rest of the family did help a little, but as Donne observed on his sickbed, "It is a faint comfort to know the worst, when the worst is *remedilesse.*"

My mother's death sentence should have been easier for me to take than Stephen's was for Jamie. The day that Stephen called Jamie, December 16, 1998, happens to be my mother's birthday. She was seventy-four that year. No one considers a natural death at that age a tragedy. But she was my mother. To her and to everyone in our family her suffering had already been a tragedy. News like that is hard to take in. Most doctors do not try to treat their own families, and maybe science writers should not expect too much from themselves, either. At first, I did not want to learn about Lewy body dementia. I did not even want to know what Lewy bodies are.

In La Jolla, as the Neurosciences Institute emptied out for the evening, Jamie sat in his office searching the Web. He scanned the abstracts of another dozen medical papers from universities as close by as San Diego and as far away as Tokyo, Tel Aviv, and Barcelona. The prognosis was roughly the same everywhere he looked.

To keep himself from panicking, he tried to focus on the positive. *In all of this bad news, it's as good as it gets,* he told himself again and again. *There is no better group than Steve's.* All of these were only averages: Individual prognoses were completely unpredictable. The cosmologist Stephen Hawking had been diagnosed with ALS when he was twenty-one. Now he was almost sixty. And at least there was riluzole, which did slow the disease at least a little. Why did it help? That might be a clue. The whole force of biomedicine was coming together everywhere. A cure might be just over the horizon of the new millennium. Maybe he could bring it closer.

Jamie found these thoughts hopeful, although a few months later, the Associated Press would run a story about ALS patients that began: "The hard part about having a fatal disease in 1999 is that the promises of miracle cures, genetic antidotes and bio-cocktails twinkle all around like lights of distant rescue ships—too far off to be of any assistance."

When Jamie looked up from his desk, four hours after Stephen's call, the sun was getting low in the sky. From the window he could see a piece of the parking lot and the playground of the institute's day-care center, empty and beginning to fill with shadows. He phoned Melinda and asked her to pick him up. When he stood to collect the papers that lay in his printer tray, he felt dizzy. He noticed, as if from a great distance, that he was shivering violently. He pulled on the warmest layer he had in the office, a wool-lined suede coat that he had bought ten years before at a secondhand shop on Newbury Street in Boston's Back Bay. Then he found his way with difficulty through the corridors of the institute and drifted outside. He felt physically injured, as if he had just staggered out of a car wreck or a collapsed building, as if he were trailing blood and going into shock. It was a mild evening but the coat did not stop the shivering. Melinda met him in the parking lot and drove him home.

They watched the sun set together from a cliff over the Pacific, chanting "Let Stephen be healed! Let Stephen be healed!" Melinda told him the sun would take their prayers around the world and deliver them to Stephen in Boston. That evening she wrote in her diary, "It is too early to be fatalistic. We don't know what is going on. We don't know. . . . Jamie is drowning his emotions in front of Quake and the TV. He played Steve tonight at Quake and talked to him on the phone at the same time. Steve, he said, sounded like he pretty much believes he's got this thing and that he's a goner. . . . Jamie said, 'Miracles do happen in our family. There is a lot of power and healing in this family.' I don't know if I can cry anymore tonight or if it's the right thing to do. It's too early, *MUCH* too early. Ridiculous to go too far when we don't know anything."

Melinda called her sister, Piper, and cried, then her father and cried. "Questions to ask ourselves," she wrote in her journal. "What do we want to be? What do we want to do? What are our options/potentials? It feels like the answer is lit for us in neon: YOU SHOULD GO TO BOSTON."

Another entry, a day later: "Peggy is whacked out and so are we.

Whacked out people should unite in support of one another. What a time this is. Sometimes I am just staring with my stomach in knots. . . . I am in a state of shock, I think. But not denial. Jamie is surfing the Net looking up ALS information, determined to come up with a cure. I am working on the miracle side of things, and almost unbelievingly telling people a little bit about what's going on."

On December 18, two days after Stephen's call, Jamie met with the institute's research director and gave notice. He said he had not found much in his year there that he could bring to the marketplace: *I don't see anything worthy of your investment or mine.* Given his family news, he had to move back home. He would be leaving the institute when his year's contract was up on February 15.

That gave Jamie a little less than two months. In that time he planned to draw on every resource he could tap in the institute. He would search all the new sciences of life. He would find some way to pull Stephen back from the dead.

Part Two

The Plan

For nothing in the whole world can

be brought

To equal the agility of thought.

Lucretius

{ N I N E }

The Repair Man

I once asked my father what he thought of Jamie Heywood's leap from engineering into genetic engineering.

"If you understand a system well, whether it is a pulley, or a circuit, or an engine, then you can deal with it," he said. "Once you understand it, then you say, OK, it's a system, and I can deal with it." An engineer could not think about life that way before the rise of molecular biology. "But today you can talk about genes, DNA, and protein. You can say, OK, I deal with systems, and as long as there is nothing in this system outside of physics and engineering, I can do it." If what has broken is nothing but a system made of molecules, an engineer can try to fix it.

My father himself had always worked on a plane much more abstract than John Heywood, author of *Internal Combustion Engine Fundamentals.* The book my father wrote while my brother and I were growing up is titled *Statistical Mechanics of Elasticity.* It gets highly mathematical by the bottom of page 2. The only part we could read was the dedication: "To Ponnie." While taking care of my mother, Dad was working on a reprint edition of *Statistical Mechanics of Elasticity* for Dover Press. The book describes the kinds of long-chain molecules you find in rubber bands, and analyzes the way they strain when you stretch one. Trying to help Dad toward the bestseller list, my brother suggested a subtitle: *Statistical Mechanics of Elasticity: And What It Can Do for You.*

Many of my father's former students were going into the new biology, where their skills with math, computers, and systems helped them study life's molecular machinery. Some of the mathematical tools my father experimented with in the 1970s and 1980s were now used routinely by big international teams of biologists and engineers. There are long-chain molecules in living flesh, too. DNA is one of them.

"That's the concept of an engineering education," he said. "You learn basic science so clear that you can then apply it to whatever you're faced with. I would say, an engineer with sufficient self-confidence would feel that way about moving into genetic engineering. Not every engineer. Many get to know a small area and never step out. But a good engineer would decide: OK, it's a system. I have to learn a lot but I can grapple with it.

"Again, this is not an ordinary fellow who would do that."

Melinda called Jamie the Repair Man. The first time they visited her mother in Athens, Jamie walked into her apartment and decided to fix the toilet. Her mother had turned off the water to the toilet tank because of a leak a few decades before. "To flush," Melinda explains in one of her many unpublished essays about life with Jamie, "you had to fill up a plastic bucket we kept in the tub, heave it over to the toilet bowl, and splash the water down in a great wave. This is how we had been dealing with elimination since I was a kid. . . ."

"Forget the Parthenon," says Melinda. Jamie spent his first day in the ancient city touring Praktiker, the Home Depot of Greece. He fixed the leak, replaced the broken seat cover, and assembled a new plastic table in the garden. Before nightfall, the American in Athens had ascended his porcelain throne, and found it good.

Unfortunately, Melinda writes, "the most appreciative audience to these bathroom heroics turned out to be the Repair Man himself."

Her mother was sure the toilet would break again as soon as Jamie was gone. She went right on flushing with buckets of water.

The first time Jamie visited Melinda's father in Manhattan, he found a different problem to fix. Phil Marsh is a folk-rock musician. In his apartment in Chelsea, according to Melinda, he had created a vast network of extension cords and phone wires, and they snaked through the long hallway from room to room. Again Jamie set off for a hardware store "to forage among wire clips and linking thingies and whatnot." He made the apartment's wiring so neat and rational that it seemed to disappear. And all Phil could do was worry about the day he might have to take it apart.

"You can see the clash of cultures for yourself," Melinda concludes. "My clan is Content With The Way Things Are, his clan Fixes Things That Are Broken."

Now Melinda was in awe of the Repair Man, and also jealous, because Jamie's search for a cure took his mind off what was torturing hers. He had vanished into his office.

"While Jamie was at work," Melinda wrote later, "I ran around La Jolla in a daze, stared cowlike at the sea." She and Jamie had lived with his parents on Mill Street when they were just starting out. Back then, Melinda had fit in beautifully in "that coven of Heywoodness," as one of her friends once told me. But now Melinda confided to her journal that moving back scared her. "A plan is needed. The couple cannot be subsumed back into the parental household, or I will lose my center and so will the couple."

On the phone, Stephen sounded calmer than anyone else in the family, and he tried to cheer her up. The day after his test at Massachusetts General, he told her, he had developed a horrendous case of hemorrhoids. He had never had hemorrhoids before, and he told her all about it.

And as I understand it, many millions of Americans . . .

Every word he said made her laugh hysterically.

Stephen told me about the hemorrhoids later on, without the comedy routines. "I know it was stress," he said. "And after that I started having problems in my feet. I would have numbness in my legs. It must have been a pinched nerve, because my whole region from my knees to my upper back was locked tight. I was taking piles of ibuprofen. I was just so tense. I mean, it hurt to move, I couldn't bend over."

He went back to the neurologists at Mass General and they put him through another MRI. Nothing was wrong—or nothing new. His right hand was still weak, and his right foot felt a little weak now.

Slowly Stephen calmed down, and his body relaxed. As his body stopped panicking, he wondered how he should spend the last few years of his life. He was in love with Wendy. On the night he learned he was dying, he had called her and blurted out, *If it is ALS, will you have my baby?* But could he really ask Wendy to do that? Would it be fair to ask her to marry him? Sometimes he thought he should buy a new Harley and take off. Out in California, one of Melinda's circus friends told her that she would love to "get Stephen pregnant." Melinda passed the word to Stephen, to make him smile.

Stephen laughed about that. Wendy was not amused. *Uh, hello*, she said. *Girlfriend standing here.*

For Jamie a clock had started ticking the instant he heard Stephen's voice in his call from Storrow Drive. Now the clock hung over him and followed him like the face of the moon wherever he went, day and night. Suddenly he felt miraculously lucky to be at the Neurosciences Institute. He had started there the very same week that Stephen went to see a doctor about his right hand. He could not be better placed to figure out how to save Stephen.

Jamie felt shy about telling his news at the institute. The year before, at his job interview, he had told Edelman about Stephen's first symptom, the weakness of the right thumb and forefinger. Now, when

Jamie told him the diagnosis, Edelman winced, nodded, and shot him a sharp bright glance of sympathy. Jamie saw that Edelman had suspected the worst from the start.

With his office mate Joe Gally, Jamie was not sure he could trust his voice. It was one thing to sit alone at his computer and search PubMed. When Jamie typed the keywords "ALS" and "prognosis," he had known that PubMed would be smart enough to search the vast scientific literature under not just "ALS" but also "amyotrophic lateral sclerosis" and "Lou Gehrig's disease." But now he had to ask Joe, the Walking Library, for the kind of research assistance that only a human being can provide.

Joe looked moved when he heard about Stephen, and he promised to teach Jamie about ALS. He warned him that the literature on the disease is confusing. ALS is a complex disease with many tributaries. In December of 1998, there were at least a dozen plausible theories about its cause. But Gally told Jamie that a remarkable paper had been published in the field that past March.

The date struck Jamie as another coincidence. It was the same month that Stephen had first gone to see the doctor.

Joe Gally had found this paper so interesting that as soon as it came out, he had introduced it at one of the weekly meetings of the institute's journal club. Joe was a very useful member of the journal club. When he brought in a paper, he could not only explain why it was exciting, he had also read all the references the authors cited at the bottom of their paper; and he knew a good deal about them, and about *their* references, too. Each piece of new science is like a brick in a city that rests on seven cities buried underneath.

The paper that interested Joe had appeared in the journal *Neuron.* The chief author was Jeffrey Rothstein, a neuroscientist at Johns Hopkins University in Baltimore who specializes in the study and treatment of ALS. Rothstein and his colleagues at Hopkins had found a defect in the machinery that nerves use to communicate. The defect was extremely subtle, and it was a kind of problem that biologists did not understand very well.

Gally had tried to explain the significance of this paper to everyone at the journal club. He told Jamie to start there.

Jamie began reading Rothstein's paper at the office. He took it home that night, along with a few more papers to help him understand it, and a few more in reference to those. Next to him in bed, Melinda was reading *Tuesdays with Morrie.* It is not a comforting book for the families of ALS sufferers—at least, not at the very start of the adventure. Melinda found it brutal to have the future brought before her face. She tortured herself with images of Stephen in a wheelchair like Morrie; Stephen with a breathing tube taped to his neck; Stephen frightened like Morrie, the old professor, terrified of choking to death, of being drowned by his own saliva. When Melinda closed the book her cheeks were wet. She had decided that she and Jamie would have to have a baby and name him Stephen, and he would play with his uncle Stephen.

Jamie heard her close the book and he glanced at her face. "I'm not reading that book," he said. He rolled over in bed with a groan. "I'm going to find a cure."

{ TEN }

An Imperial Message

Kafka wrote a short short story that might almost be about ALS. It is called "An Imperial Message."

The emperor, so a parable runs, has sent a message to you, the humble subject, the insignificant shadow cowering in the remotest distance before the imperial sun; the emperor from his deathbed has sent a message to you alone. . . . The messenger immediately sets out on his journey; a powerful, an indefatigable man; now pushing with his right arm, now with his left, he cleaves a way for himself through the throng; if he encounters resistance he points to his breast, where the symbol of the sun glitters; the way is made easier for him than it would be for any other man. . . . But how vainly does he wear out his strength; still he is only making his way through the chambers of the innermost palace; never will he get to the end of them; and if he succeeded in that nothing would be gained; he must next fight his way down the stair; and if he succeeded in that nothing would be gained; the courts would still have to be crossed; and after the courts the second outer palace;

> and once more stairs and courts; . . . and if at last he should burst through the outermost gate—but never, never can that happen—the imperial capital would lie before him, the center of the world, crammed to bursting with its own sediment. Nobody could fight his way through here even with a message from a dead man. But you sit at your window when evening falls and dream it to yourself.

Stephen's brain was sending messages continually to the muscles of his body. If he had been as healthy as he still looked and felt, those messages would have traveled in a flash from his brain all the way down to his toes, or to the thenar eminence of his right hand. But the brain's messages travel in stages. To reach the limbs they have to travel out through nerves that run through the gates of the skull and down through the spine. Those are the motor neurons. They are bundled together like cables, and each individual nerve is a long, thin, delicate strand. These motor nerves carry all the messages that command the muscles in the limbs to work: to run, walk, hammer, saw, turn a key in a lock.

Stephen's motor neurons had begun to die, one by one, and messages were failing to get through to his right hand. As more of those nerves died, fewer and fewer messages would get through the spine, and the rest of Stephen's musculature would wither away.

Before he read Rothstein's paper, Jamie Heywood dreamed of a cure, but he had no plan of attack. Mostly he was trying to become an educated consumer. He called his brother and his mother in Newtonville a few times a day with advice about vitamins and the drug riluzole. Now, as he absorbed Rothstein's hypothesis, Jamie began to develop a maddening hope that he really might save Stephen.

Rothstein looked at the molecular action inside the cell, and there he told a story that is like Kafka's parable writ small.

The life of the cell is almost as complicated as the life of a city, and the cell's architecture resembles a medieval city's. The cell has outer walls, and somewhere deep inside the walls are the walls of its dark and massive nucleus. Walls within walls, like the Castle in old Prague, the city and inspiration of Kafka.

The nucleus is packed with DNA. This DNA is powerless to do anything by itself, just as the brain is powerless without the body, or an emperor is powerless without the help of a thousand servants, minions, and messengers. Messengers are continually streaming in through the gates of the nucleus. These messengers bring word from the very edges of the cell and from outside its walls. Other messengers are continually streaming back out through the palace gates bearing replies that must travel to the city walls and beyond.

These messages are as vital to the life of the cell as the messages of the brain are to the body. Outside the walls of each nerve cell, for instance, there is always a standing pool of chemicals, including glutamate. This pool lies in the spaces between nerve cells like water and mist in a moat. If glutamate levels in the pool begin to rise outside the nerve cell walls, the cell goes into crisis mode, much as if the River Elbe were rising in Dresden. All along the moat a corps of engineers, really just a crew of glorified bailers, fights the rising tide before it floods the city.

These bailers are molecular machines. Sometimes they wear out. Sometimes in the rising tide they begin to be overwhelmed. Then chains of messengers rush from the outer wall through the old city squares and into the palace: *We need bailers!* And the emperor replies, *Conscript bailers!* If this imperial message should fail to reach the city wall, the bailers would be overwhelmed. The flood would rise. The city of the cell would die.

In Rothstein's paper, he makes his way into the very heart of the cell. There he spells out precisely what the emperor whispers into the ear of the imperial messengers, and what the messengers take away.

In the first days of January 1999, about two weeks after Stephen's call, Jamie asked Joe Gally to help him prepare a list of questions for Rothstein. Then he steeled himself and phoned Rothstein's office number at Johns Hopkins. Jamie was thrilled and surprised when Rothstein actually took the call. Sitting in his monk's cell at the institute, Jamie went through his list of questions one by one, scribbled pages of notes, desperately trying to keep up, and trying not lose his list in the growing pile of paper. Then he spent an hour and a half going over the notes with Joe Gally.

To understand Rothstein's argument about ALS, Jamie had to know a little more about the way all this messaging works inside the cell. The DNA inside the nucleus is so imperially helpless that by itself it cannot even read its own messages. A fantastic army of servants inside the nucleus helps the DNA to read each one. This act requires, in practice, the unfolding and unpacking of one small portion of the DNA's long, carefully wound scroll to expose just the one spot on the DNA that can read and respond to that particular message: in this case the message that says, *Need a bailer out here.*

DNA cannot read the message by itself and neither can it reply. By itself it is as helpless as a brain. Its army of servants has to work together to cobble together a reply. These writers, like the bailers out at the wall, are highly specialized molecular machines. As a matter of fact, DNA by itself is so helpless that the metaphor of the emperor at the center of things may someday have to go. The workings of a cell require so many hands, each of which is indispensable to life, that the cell is really more like a democracy than a kingdom. It is only because of our deep-seated romance with monarchs, our taste for the glamour and trappings of royalty, that we like to think of DNA as the imperial molecule. Biologists with leftist leanings make this point all the time, passionately, and they are correct. Every working part of the cell is vital, from the giant coils of DNA at the center to the tiny bailers at the moat.

In human DNA, the imperial message "conscript more bailers" requires a long message in the genetic code. Some of the letters in the message are *gctcaccccg gcgtccgctt tctccctcgc ccacatctgc cggatagtgc tgaagaggag*. A few more are *aaggatatga gtctcagcaa attcttgaat aaactcccca gcg-*

tatccta tggta. Now, contrary to the view that molecular biologists developed in the early days of their science, this is not the only information that dictates the message that goes out. DNA's metaphorical servants and attendants have some freedom, some wiggle room, in transcribing those letters and assembling the message that will go back out through the gates. They have to work with the message in the DNA, but they can splice it together in many different ways. The message they splice together is made of a slightly different molecule called RNA. The spliced message that must now make its way out through the wall of the nucleus is called messenger RNA.

Now we are deep in the heart of the palace of life, at the moment when the dying emperor whispers to the messenger. What Rothstein had discovered was that in many of his patients with ALS, the alternatively spliced messages were failing. The cell had to make more bailers or it would die, and then the whole body and all its cells would die; but the messages were bad.

In the cell, messenger RNA makes its way out into the crowded throngs outside the palace walls. There it finds its way to a piece of heavy machinery called a ribosome, where it is translated. The ribosome travels along the messenger RNA, reading the message and constructing a large ribbon of a molecule. This is a protein.

Now as the protein emerges from the ribosome, it carries a sort of imperial signal, and it runs through the cell toward the wall, as in the race of an imperial messenger. If the protein emerges from the ribosome without that imperial signal, it can go adrift and never reach the cell membrane. There are many diseases that occur at that stage, and they are not all neurological. The best known is cystic fibrosis, in which a simple protein emerges without the imperial signal and fails to get exported through the cell membrane. For lack of that one protein, the body's lungs slowly become coated with slime, until the body drowns.

Technically, when the message in the DNA is turned into messenger RNA, the act is called transcription. When the messenger RNA is used to build a protein, the act is called translation. But essentially we are talking about a chain of messengers in which something can go wrong at each point.

Even the folding up of the ribbon of protein into its proper shape is a kind of translation, in a sense. And if the proteins get misfolded, they can cause a long list of diseases. And that is what goes wrong in ALS, Rothstein argued. The misspliced RNA led to mismade proteins and they never made it out of the cell—or they failed at the wall.

Glutamate is always lying in pools outside each nerve cell because it is the most common chemical compound that nerve cells throw at each other in order to pass messages back and forth. (Other messengers are serotonin and dopamine.) Glutamate is an ancient substance in the body. It is one of the body's amino acids, the basic building blocks of proteins. If the nerve cell cannot get rid of the glutamate as fast as it piles up, then the glutamate overstimulates the cell. The cell does not literally drown in glutamate; it gets overexcited. It works itself to death. This is called excitotoxicity.

The bailer has to transport glutamate out of the moat to points inside a nearby cell where it can be safely dismantled and recycled and used to make more proteins. The bailer in question, the crucial bailer that Rothstein was focusing on, is a protein called excitatory amino acid transporter two, or EAAT2, pronounced Eat-Two.

At the Neurosciences Institute, Jamie pieced all this together as fast as he could. Stephen's nerves were dying, so they were not sending signals to his muscles. For Stephen the first sign of trouble had been the thenar eminence. The muscle in there had not been getting signals from Stephen's brain because the nerves in his spine that carried the message to that muscle were dying. And so the muscle had withered away, and a gifted young craftsman had lost the use of his right hand just as he figured out what his hand was made to do.

If Rothstein was right, the problem had started with the splicing of the imperial message, which caused the failure of those tiny bailers at the cell's outer wall.

It was so complicated, and yet it was also simple. If Jamie could get the correct message into the nuclei of Stephen's dying nerves, then the message might get out through the palace walls and race through the city of the cell to its outermost wall, and then far, far beyond, until it found its way at last into those waiting hands.

{ ELEVEN }

Hope and Science

Jeffrey Rothstein and Robert Brown are physician-researchers: scientists who do medical research and treat patients at the same time. There are many life scientists who never see patients. Molecular medicine more than any other rests on the studies of a vast number of biologists who are essentially doing the work of anatomists, but within the cell, tracing the structures and functions of the almost endless varieties of molecules that perform the dance of life.

Rothstein's paper about EAAT2, for instance, led into two kinds of unknown territory. One was alternative splicing. No one really understood how the cell manages to do the work of reading DNA and transcribing its message into RNA inside the nucleus.

The other rough, dark territory was protein transport. The EAAT2 gene made a protein that did not get transported properly when the DNA message was mistranslated. When the protein did get transported properly, it worked to transport another molecule, glutamate. Protein transport inside cells is still poorly understood. It takes endlessly patient and hopeful work to understand.

One of the pioneers in that field is Günter Blobel. In English, depending on how one pronounces the name, it can rhyme either with "noble" or with "Nobel." That year, Blobel would become Rockefeller's twentieth Nobel Prize–winner. His story is as much a quest as Jamie's, but the older man came out of pure science and operated on that very different timescale, the urgencies of a clock whose hands take lifetimes to go around once.

It began in early February 1945, which happens to have been the same moment that scientists at Rockefeller were making the first electron micrographs of the architecture of the cell. Blobel was a child of German refugees. They came from a village in Silesia, they were fleeing the Russian front, and they were passing through Dresden, the city that called itself "Florence on the Elbe." Günter Blobel was eight years old and it was the first city that he had ever seen. He had always loved architecture, beautiful spaces, beautiful buildings. Even at six or seven, he would ask his parents if he could go into a church. They would give him a polite, baffled shrug. *You go ahead, we'll wait for you out here.*

In Dresden he saw palaces, castles, churches, and the city's famous cathedral, the Frauenkirche, Church of Our Lady, which floated over the baroque city, a great stone dome in the shape of a bell. The war was almost over. Firebombing had already destroyed most of Berlin, Hamburg, Cologne, on and on. But in Dresden the only treasure the Germans had lost they had burned themselves. On November 9, 1938, which the Nazis called *Kristallnacht,* the Night of Broken Glass, they had torched the city's synagogue, which had been designed by the same architect who built their opera house.

The Blobels made their way to a relative's farm twenty-five miles to the west. On the night of February 13, 1945, a huge armada of Allied planes flew low over the village, dropping flares. Half an hour later, the sky to the east, toward Dresden, began to glow red with a false dawn. Soon the light from the fire was so bright that standing outside the blacked-out farmhouse they could have read a newspaper.

That June, the boy was in a wagon train of refugees that wound its way east and crossed the valley of stones that had been Dresden. Even the Frauenkirche was gone in the plain of smoking stones. The boy smelled a stench that was heavy and sweet and bitter, something like the smell of cherries. The wagon train reached the Soviet border and turned back. Not long afterward the boy and his family passed again through the desert of stones, now with everything they owned in a wheelbarrow, and took shelter once again on the farm to the west of what had been Dresden.

When he grew up, Blobel found his way to the laboratory at Rockefeller where they had opened up the baroque architecture of the cell. He built on the work that was begun there, and arrived at a bold young man's idea that he called the signal hypothesis. Blobel was a scientist's scientist, someone doing pure research that seemed very far from medicine, research for the pure joy of understanding. He had the same joy that Lucretius, the poet of science, sang in ancient Rome: "There is more to learn; for me it's pleasant work." "I feel a more than mortal pleasure in all this."

He spent thirty years proving the signal hypothesis. He proved it essentially by disassembling and reassembling the cell's working parts in test tubes. He engineered exact facsimiles, restorations, of the beautiful architecture of the signal and of the invisible halls and doors through which it has to pass from the nucleus to the outer membrane. Like Kafka's imperial messenger, any protein that was bound for the construction and reconstruction work inside the cell had to carry a signal. Blobel found the signal. And he saw that if the messenger did not have just the right signal, like the medallion worn by Kafka's messenger, the message could not get through the cell. Sometimes it could not even get through the pores of the nuclear membrane. "But how vainly does he wear out his strength; still he is only making his way through the chambers of the innermost palace. . . ."

For thirty years, Blobel studied the ways a cell transports proteins. It was like getting to know an ancient living city whose restoration work is never done. In his mind, Blobel compared the masterpieces of molecular architecture that he was discovering to the wonders that he remembered from the beautiful doomed city of Dresden. The baroque architecture of the nuclear pore reminded him of the Frauenkirche, Church of Our Lady, although his students and colleagues laughed when he said so. When he won the Nobel Prize in 1999, Günter Blobel held a press conference at Rockefeller and announced that he was donating his prize money, almost a million dollars, to the restoration of the church and the synagogue of Dresden.

{ TWELVE }

Almost No Time

In the first weeks of January 1999, Jamie sat at his beautiful redwood desk with the green, architect-designed file drawers underneath it. It was still covered with neat piles of papers about the different projects he had worked on for the Neurosciences Institute. Those piles were steadily disappearing under papers about ALS and stacks of medical books and guides, including, still, *The Human Brain Coloring Book.* After a lifetime of dyslexia, suddenly Jamie could read with ease—and read at enormous speed. On his computer's nineteen-inch monitor, he kept many browser windows open at the same time. He would follow a lead from one of Rothstein's papers and check it out on PubMed. Then he would click on "Related Articles" and scan a list of twenty abstracts. With a click he opened each of them in a new window. Then he paged through and discarded some and copied the ones he wanted into his files. "I could read an abstract in just—*flash,*" Jamie says. "I mean, I was like, *boom, boom, boom, boom.* My mind was in some kind of warp or something. I would click, and then I'd go, *That's relevant, not relevant; relevant, not relevant.* I was already very fast at using computers. And I remember just sort of flying through things."

Jamie cannot explain why his dyslexia vanished. "You know, everyone laughs at me when I say this, because I went to MIT," he says. "But I'd *never learned anything by reading before.* And all of a sudden—you know, I could have gone through medical school! I could read, I could understand, I could remember. Obviously I was surrounded by astounding teachers. But it was unreal, this period of time."

He left the institute often now and went to conferences nearby at the Salk Institute and the University of California at San Diego, lectures on anything that might be relevant to his search: mad cow disease, Alzheimer's, Parkinson's. He visited the offices of local ALS specialists and clinicians, most of whom had the same plastic model on display, a curving human spine and skull. He made appointments with the lecturers he thought might be able to tell him something; he talked with neurobiologists, geneticists, molecular biologists.

Around the institute, Jamie began to act as obsessive as Joe Gally. He skipped sleep, meals, haircuts. He wandered among the institute's futuristic buildings and stared at the pebbles in the paths. Normally he walked very fast when he was going somewhere. Now he moved deliberately, looking down—not staring at the ground as if he were trying to avert his eyes from the world, but as if he were searching for something there in the stones. He even drove slowly. After a day of tutoring sessions with Joe, and confusing sessions in lecture halls, he drove home trying to calculate how much time he had. This was hard to do, because the course of the disease is so variable. How would it take Stephen? When he wondered about this, his guesses were tinted by his family pride. He was sure that with a strong young carpenter like Stephen, the disease would go slower than with most.

And how would Stephen take the disease? That mattered. What would he do? The show that *60 Minutes* had run about Doctor Kevorkian, the show in which he killed a man with ALS on the air, was still in the news. It was one of the most controversial shows in the program's history. Many terminally ill patients were outraged at the implication that life with advanced ALS or Parkinson's was no longer worth living. The producers did a follow-up show about ALS patients with a passion for life.

Jamie was sure that Stephen would want to live. But he could not imagine how Stephen would feel when he could no longer do what he wanted to do. How would he react when the reality set in? It had not sunk in yet and it might not for a while. Stephen still felt fine, almost immortal. Jamie once took Stephen to meet a musician named Steven Fowler, who had an advanced case of ALS. Fowler was a skeleton in a wheelchair, rocking back and forth slightly to help his breathing machine force air in and out of his lungs. Stephen was full of pity, but he could not believe that he would be reduced to that himself. When Stephen was out of earshot, the musician whispered to Jamie, "He's in denial."

Jamie knew that Stephen might have only a year or two before he was in a wheelchair. To save him while he was still walking, Jamie had to start treating him within one year: He had to start early in 2000. To start then, he had to persuade ALS specialists to test a new approach by the end of 1999. To do that, he had to find an idea right away. He had almost no time at all.

"I really started having some hard sessions with Joe," Jamie says. Gally would coach Jamie in all of the fields that radiated outward from Rothstein's paper about EAAT2—genetics, neurobiology, neuropharmacology, neurogenetics—and he would help Jamie prepare lists of questions to ask specialists. "I would spend an hour on the phone with someone," Jamie says, "and afterward we would spend an hour and a half debugging my notes, and we would do it again, and we would do it again. And there must have been twenty conversations like that."

Jamie still has a sketch that they made after one of those talks, a rapid technical drawing of a section of human spine, drawn half in his own hand and half in Joe's.

The information he was learning with Joe was coming together in his head in graphic form, an indescribable, three-dimensional pulsing matrix of elements that hit and pushed and brushed against each other day and night like nerves in a network or like molecules in a synapse. Not only was his dyslexia gone, but overnight he had acquired a photographic memory, just like Joe Gally's. And the information he was learning kept coming together in his vision.

"I've never had anything like that in my life," Jamie says. "I was daily listening to myself talk and not believing what I was saying.

Between talking to the doctors and then talking to Joe, somehow I just learned what kind of data was important and what wasn't. Like the needle of a compass."

Sometimes now he remembered the hero of his favorite book, *Ender's Game* by Orson Scott Card. He remembered how Ender felt when he had been flown up into a space station to be trained on an emergency basis in how to defeat the enemies of the human race, the Buggers. "So Ender withdrew into his studies and learned quickly and well," Card writes. "Whenever he was given a problem that involved patterns in space and time, he found that his intuition was more reliable than his calculation—he often saw at once a solution that he could only prove after minutes or hours of manipulating numbers."

But in the evenings Jamie and Melinda kept seeing Stephen in a wheelchair, Stephen with a plastic breathing tube in his trachea, images that brought on boiling nausea. To block out their visions of Stephen, Jamie and Melinda roller-skated themselves to exhaustion in Mission Bay, which sometimes helped. Then Melinda belly danced at a club—set after set, night after night. Jamie sat alone at a table for two and watched Melinda. At the institute, Jamie had talked a little about his crisis with a friendly Italian psychiatrist-turned-neurobiologist. *So, Jamie, do you drink?* the man asked kindly. *A little. Good. Your parents? A little. Good. You're going to need a few.* Melinda danced in her costumes of dangling coins, and Jamie drank vodka and tonics, sometimes more than a few.

No matter how wildly they lived, they could not escape their dread. Their digestive systems were in revolt. "Biography becomes your biology," Melinda wrote in her journal. "Every conscious thought generates a physiological response. It's driving me mad. Everything is driving me mad. I'm mad!"

Although Jamie was hardly sleeping now, on some days he still went racing in San Diego Bay. Water had always moved him spiritually: the surf at Duck, Boston Harbor, even the smoothness of the Charles River and the lagoon in the Boston Garden. Racing was awesome. That had not changed. It took only a little of his strength, but all of his attention. Racing was one of the few things that could make him stop thinking about ALS.

Once, at Duck, the Heywood boys and their friends and cousins were playing basketball while the elders sat and drank daiquiris and cheered them on. In the middle of the game, Jamie saw a young man drifting too far out on a windsurfer. It was a Costa Rican exchange student. A cousin of theirs had brought him to Duck that summer. His windsurfer was broken, and he was drifting toward Costa Rica. It was Jamie who got on his surfboard, went out, and brought him back. The man was not in real danger yet and nobody else had been moved to do anything about it. Now sometimes when Jamie was looking out at the Pacific, he felt his brother drifting away.

Jamie and Melinda had spent that Christmas in Phoenix with Circus Flora. Back in Newtonville, John and Peggy had celebrated the holiday with Stephen, Wendy, and Ben. What a difference from the Christmas before, when the Heywoods were helping Stephen finish his house in Palo Alto. Stephen had not been diagnosed officially yet by Doctor Brown; the appointment was still a few weeks away. But they knew Stephen had ALS, and the thought of it, daily, nightly, hourly, hit them literally in their guts. They tried to bear the news and the suspense quietly in each other's company. They were not people who let themselves wail out loud.

Peggy and John saw no point in keeping the news from their family and friends. They each write an annual Christmas letter. Peggy sat down at her computer with its desktop picture of her three boys digging in the sand at Duck. "I can't even begin to describe what it has been like," she wrote, "except to say that I now realize that all the metaphors of grief have their physical equivalents. I am thankful for the times when I don't feel awful and for when I don't wake up in the middle of the night. Stephen has been his usual low-key sweet self, and is able better than I to say, 'We'll see what the doctor has to say in January' . . . and I do pray for the best. The three boys are so close to each other, as is Melinda, my daughter-in-law, so it has hit us all this Christmas fiercely."

John did not write his own Christmas letter until January. "A little later than usual, but you will see why below. As Peggy said in her end-of-year letter, it has been the 'best and worst of years.' Stephen is perhaps doing better than the rest of us—steady and low-keyed. As our friends and church community have learned about this they have been incredibly comforting and supportive. We will keep in touch as we learn more and things develop. I'm sure we will be spending as much time together as a family as we can."

Stephen himself was working in slow motion on the house on Mill Street. He was laying tile and yellow limestone in the new master bathroom. He felt fine, except for his claw hand. He was also having some slight trouble now with his left foot.

People can react in several ways to a diagnosis of ALS. They can devote themselves to fighting the disease, they can close one eye to it and try to live as fully as they can, or they can give up—not an option with the Heywoods. Jamie was masterminding a war, and Stephen was calling friends from Alaska to Corfu to tell them his news and make plans to get together. They all heard him on the phone with his voice as dry, cheerful, and sardonic as ever: *Yeah, I'm toast!* They could not understand how he could keep his good nature. But good nature is strangely common in victims of ALS. Doctors often say that ALS patients are some of the most painful patients to treat and lose. Surprisingly often, they are sturdy, athletic, sunny optimists, like Stephen. (Other diseases prey on couch-potato pessimists.) That year, ALS advocates would parade in a fund-raising demonstration on Capitol Hill—walking or in wheelchairs—wearing T-shirts that read, "ALS patients are such nice people, it's a shame they'll only be here two years."

After the shock of the news, one of the first things Stephen discovered was that he wanted to have a child. "I don't want this to get me in trouble, because my family loves Grace Church—but I'm an atheist," Stephen explained to me a little later on. He said this with a small smile, because his family's faith was complicated. John and Peggy were both unbelievers, or so they claimed, even though they went faithfuly each Sunday. "I'm so atheist," Stephen said, "and have been since I was twelve or eleven. So if you're atheist and you know you're gonna die before you get so exhausted from life that you don't really care, you're kind of fucked, because there's nothing to look forward to. And one of the first things I thought about was kids. Well, maybe I would feel differently if I had left behind a whole bunch of buildings, too. But kids—that's what you can leave behind."

Even though he and Wendy were not even talking about marriage, Stephen found himself thinking more and more about having a child with her. She was thirty-four. He knew she wanted a family. One of her older sisters had chosen artificial insemination at about her age. That circus friend of Melinda's had joked about getting Stephen pregnant. Now he thought: *Wendy.*

Those who loved Stephen wept when they listened to him, knowing how short a time he had left. "I mean, my mom was bawling," he told me. "But Wendy and I started talking about babies. Wendy would make a great mom. I mean, shit, I could go out and plant my seed if I wanted to, but you know, how do I know what's gonna happen after I'm done? How can I trust to that? And I felt I could trust Wendy. So it was very intense and very strange.

"It's atrociously bad timing, all of this, I guess. It's a very hard decision. I knew I was in love back in December, before any of this shit happened, which is great. But normally you'd just be like, 'Wooh-hooh, we're in love! And now we can just hang out and go to the movies and shit like that, and maybe we'll live together and maybe we won't.' " Now almost nothing was normal. On the one hand, Stephen and Wendy were talking about a baby. On the other hand, they did not even want to move in together without being engaged. Stephen had done that before, with Stephanie, and it had been a mistake. "Everybody knows that if you live together before you get married, it's not gonna work."

So there he was, in love and dying. Stephen knew that Wendy loved him, too, and what was the right or fair thing to do? "I mean basically Wendy's fucked, you know. No matter how this works out, it's not good."

{ THIRTEEN }

Prisoners in Outer Space

On January 13, 1999, Stephen went back to Massachusetts General to receive the official diagnosis from his neurologist, Robert Brown. Normally he preferred to go to doctors' appointments alone. But his mother wanted to come to this one. Jamie asked Stephen if he could meet Doctor Brown, and Stephen decided to let him come, too. So Jamie flew home again, with his multilayered vision of molecules still forming in his head.

John Heywood did not ask to come, and Stephen understood. John bicycled to the Sloan Automotive Laboratory that morning, just as usual.

Bob Brown has been working on ALS for more than a quarter of a century, and he is one of the most distinguished scientists in the field. In 1993, Brown and his colleagues discovered that a small fraction of ALS patients inherit a mutation in a gene called SOD. Brown's discovery enabled ALS researchers to genetically engineer a mouse with a bad SOD gene. For the first time, scientists could study the course of the disease in a laboratory animal. Normal mice live two years. ALS mice die at about five months, paralyzed and choking.

The creation of an animal that suffered and died like that was a miserable thing for the mice, but it brought hope to any human being concerned about ALS and the other nerve-death diseases. It allowed Brown and many others to experiment and try to figure out what is going wrong and how to fix it, although no one had found a cure. In medicine the easy diseases to cure were already curable. As a colleague of Brown's, another Harvard neurologist, once put it, "The low-hanging fruit has been picked."

It is one of the many coincidences in Stephen's story that he found his way to Brown. Stephen had not known that he had ALS when he made his first appointment. Bob Brown is white-haired, patrician, and, for a Harvard man, extremely self-effacing. Delivering a diagnosis of motor-neuron disease is equivalent to handing down a death sentence. As a world-renowned authority on the disease, he has to deliver it at least once a day. He always hates it. The mice in his laboratory never know what is happening to them. But most of Brown's patients, people in their fifties and sixties, feel as if they have a long time to live. Some of them are young and vigorous, like Stephen. Bob Brown is a kind man and he has never gotten used to any of it.

Less than a month had passed since Stephen's neurological workup on December 16. In the old days, doctors did not even give patients a diagnosis of ALS at this early stage. Since there was nothing they could do for them anyway, they kept the bad news to themselves for a while. But that was before the ALS drug riluzole. Now that doctors could offer even a few months' extension of the death sentence, they told their patients the diagnosis promptly.

Brown's main goal in that first appointment was to acclimate his patient to the diagnosis. He knew where he was going and so did the Heywoods, but they all found it hard to take the steps one by one that would lead them there.

Brown's office is a cubicle with just enough room for a desk and a small curtained alcove. Stephen and his mother sat down in chairs facing the desk. Jamie pulled in a chair from outside and sat by the door. The room was so small that both brothers could hold Peggy's hands while they waited for Brown to begin.

Jamie had brought a tape recorder so that he could listen to

Brown's diagnosis again later. But when he pushed the button he discovered that he had forgotten to bring a tape. Before they started, he borrowed one from Brown.

"Any changes since December?" Brown began in a doctor's studied, friendly, professional voice.

Stephen told Brown that he thought he noticed a little weakness in his left foot.

"Mmm-hmmm."

"And in my right foot it's much more pronounced."

"Ah-huh."

"Other than that, no, not really any changes."

"OK."

"The last couple of weeks have sort of been strange because we had Christmas, and, uh, I've had these stomach problems that have been very painful," Stephen said. He meant the hemorrhoids, and also a case of stomach flu. "So I have felt a little under the weather in general. But, you know, you start looking at everything and wondering if that's related, and you get a little paranoid."

"Sure."

"So I think I've just had a bad couple of weeks," Stephen said, with a chuckle. His tone was eagerly manly, as if to say, *I can talk about this straight, no big deal, I can handle all this.*

"OK," said Doctor Brown. "So you're essentially functioning." The phone rang. He dealt with the call, and turned back to Stephen. "Your voice sounds normal to me, it's been fine?"

"Yeah."

"Great. And chewing and swallowing are fine?"

"Fine."

"OK, good." He asked Stephen about the stomach problems, and gave him a little advice about the hemorrhoids.

"Alrighty. Well, let me examine you. And then we can sit down and talk a little."

So I can try to summarize what we have at this point," said Doctor Brown, after the exam was over and he and Stephen had settled themselves again on opposite sides of his desk. "And then I'll of course try to answer your questions."

"OK."

"You have a history of motor weakness, which started very slowly and focally and has kind of slowly increased a little bit in its distribution and certainly in its severity.

"The EMG testing that you've had showed focal motor-nerve abnormalities at first, and now shows more widespread abnormalities, which involve the arms and the legs, as well as the initial site of onset. Not the face or the tongue.

"Um, and so that's the picture of a progressive motor-neuron problem. And, um, in addition, ah, your examination shows, ah, some suggestion that what we call upper motor neurons that reside in the brain and affect the motor neurons that reside in the spinal cord, also may not be functioning quite normally. And so—"

Just as Doctor Brown steeled himself to deliver the death sentence, Jamie interrupted from the doorway.

"In the cortex?" he asked.

"Yeah," said Doctor Brown. "I'll draw you a picture of that." He got out a pen and a legal pad and they all gathered to look. "So there's a suggestion then that you have involvement of both upper and lower motor neurons," Doctor Brown said. "So, for example, if this is, let's say, your brain, and this is the base of the brain, and this is, let's say, the spinal cord down here, this is—apologies to those of you who are *real* artists—a stick figure of you." The doctor chuckled apologetically. "What we're saying is that there are motor neurons that live in the motor cortex and descend to different levels of the base of the brain and the spinal cord. These are so-called upper motor neurons. And then there are others that reside down at the base of the brain and go up to muscles—for example, of chewing and swallowing and of the tongue. Other nerves reside in the neck and go down to muscles that have to do with arm and hand movement. And then still others that live elsewhere and go to other muscles down further in the body.

"So what were saying is that, um, there's clear electrical evidence of involvement of the lower motor neurons—"

"Uh-huh," said Stephen.

"Fairly widely in the body. And then there's a suggestion of briskness of reflexes, which is a token of abnormalities of the upper motor neurons. So that points to some dysfunction up here, too."

"So you're sort of saying they're highly primed," said Jamie.

"I'm saying the reflexes are very brisk," said Doctor Brown, with a faint testiness in his voice. Jamie kept breaking in just as he was about to give Stephen the diagnosis. But that was all right, of course. Jamie had a right to ask. "Yeah," the doctor said, more patiently. "And that often correlates with trouble in the upper motor neurons."

Stephen could see that Brown was about to break the bad news. He felt intensely alert. He could read it all in Brown's face. John Donne wrote on his sickbed, "I observe the *Phisician,* with the same diligence, as hee the *disease;* I see he *feares,* and I feare with him: I overtake him, I overrun him in his feare, and I go the faster, because he makes his pace slow; I feare the more, because he disguises his fear, and I see it with the more sharpnese, because hee would not have me see it."

Nothing has changed since Elizabethan times, the terrible theater of two, or three, or four, in which things are spoken aloud as if for an audience, with formality, as if forever, the doctor proceeding in a slightly stilted, formal voice, almost like a judge at a sentencing, with no condemnation, of course, but with all the finality of a hanging sentence. It is one of the most private moments in a life and it is also somehow like a piece of public theater. The doctor, the patient, and the family are marshaling the evidence almost ceremonially, in a ritual, to announce what they all dread to speak and hear, the words that, as they are pronounced, will condemn one man to death.

Stephen felt acutely aware of it all. Even his doctor's use of all those spacers, the ums and ahs, were deliberate. They were for kindness, to signal that something big is coming and to let it come a little more slowly.

"So, um, you know," said Doctor Brown, "the, uh, diagnosis, uh, that unfortunately—that all of this points toward—is motor-neuron disease, or it's Lou Gehrig's disease, or all these terrible terms in ALS. Um. Our job is to work *with* you to do everything we can humanly do. First of all, to see if we can find something treatable here, and then to find a way to manage the problem if that's the corner we're painted into."

"OK," said Stephen.

"And, uh, you've had a pretty compulsive set of studies," Doctor Brown said, meaning comprehensive. He walked Stephen step-by-step through all the tests that he had now undergone, and listed all of the lesser medical conditions the tests had ruled out. Compared with ALS, even brain cancer is a lesser condition.

Stephen followed every syllable of Brown's speech, nodding to show that he knew, he understood, that he had expected this. Later on when Melinda listened to the tape and transcribed it for Jamie, it broke her heart to hear Stephen's strong voice, his effort to say with every breath, *I'm all right.*

Doctor Brown drew a few more diagrams of Stephen's brain and nerves, to explain everything the tests had ruled out. Stephen kept repeating that he understood. At last Brown said, "I think the real question is where to go with a diagnosis that I'm afraid we're not going to be able to escape."

"Yes," Jamie breathed from the doorway. He glanced at Peggy. She looked as if she were sitting in a prison cell in outer space.

When Doctor Brown had finished talking, Jamie could not help broaching the subject of EAAT2, although he could tell as soon as he began that it might be the wrong moment.

Stephen listened. He knew from long experience that he was hearing the start of one of Jamie's adventures, although he had no way to guess then how far this adventure would go. After a few minutes of Jamie's gibberish—*Eat Two, Eat Two, Eat Two*—he decided that it was time to wrap things up. He had to get Jamie back under control.

("I do that with Jamie a lot," Stephen told me later. "Which I shouldn't always do.")

"So this afternoon I think I'm going to have to go buy a Harley!" Stephen said. It was not much of a joke, but they all laughed anyway. Then Stephen said, "I think Jamie also wants to know whether it's appropriate to call when he has occasional questions."

"Sure," Doctor Brown said kindly. "Sure."

"Because he has a much better theoretical understanding of things than I do," Stephen said. "And I'm not necessarily going to *plan* on having a theoretical understanding of this stuff," he went on. "Because I don't—"

"He's very good at delegating!" Jamie put in from the doorway.

But Stephen refused to let Jamie interrupt again. "Because I don't think it's critical to do so," he concluded, with a soldier's dignity.

Doctor Brown had already ruled out Parkinson's disease and stroke. But to make absolutely sure that his diagnosis was correct, he recommended one more MRI. The two brothers and their mother had lunch near Doctor Brown's office, where Jamie could not quite keep himself from crying. Then they went for the MRI. There was one other family in the waiting room, and the parents were hysterical because their child might have a brain tumor. But Jamie and Peggy chanted softly to each other, "Brain cancer, brain cancer, brain cancer." With cancer at least there was some chance of a cure. With ALS there was no chance at all.

That evening back in Newtonville, in the old house on Mill Street, after his parents had gone to sleep upstairs and Stephen had gone down to the Cave, Jamie sat up late with an old friend of his from high school and MIT, Bjoren Davis, a software engineer. Back at MIT the two of them had worked together in one of Jamie's first entrepreneurial efforts, Heywood Associates. They had designed Amtrak's train arrival and departure display at Boston's Back Bay Station, and it was still in use.

They drank beer in the solarium with the dark sky and the bare trees overhead in the glass ceiling. Jamie told Bjoren everything he knew about the EAAT2 protein. The one key element that every ALS patient shared was that this EAAT2 protein was missing.

Bjoren looked at Jamie. He said, *Well, why don't you just put the damn protein back?*

At that moment, Jamie says, he began to lose his photographic memory and began to think like an engineer again. For the first time, he began to feel the can-do techie spirit returning. He saw exactly what he had to do. It was not so different from what he had been doing all along since MIT, putting together teams to do a job. Only now he needed a dream team, and he had almost no time.

On January 16, Jamie Heywood traveled for twenty-four hours to get back to California. Plane after plane was delayed or rerouted. In Atlanta and again in Cincinnati he slept on airport floors.

Melinda wrote in her journal: "Jamie sobbed and sobbed when he got home, wept like I've never seen before, overtaken by the emotions he had kept in check all week in Boston. He shook and heaved, and my tears came to join the party. Stephen has some rule that says no two people can cry at once in his presence but he wasn't there to police us.

"Then Jamie took a shower and went to work."

{ FOURTEEN }

A Forced Journey

I was not in on the Heywoods' story from the beginning. Late one morning in February 1999, just after Jamie got back from Boston, Ralph Greenspan called me from the Neurosciences Institute.

"Have I got a story for you!"

"That would make two," I said.

Ralph Greenspan had led me to the book that I had just finished writing, *Time, Love, Memory,* which is about genes and behavior. The book was on its way to the printers that day. I had a dummy of the jacket propped up on my desk where I could stare at it.

"This one would be a little more poignant than the last," Ralph said. "And also a little more compelling."

It was then that Ralph told me the story of Jamie Heywood, the brilliant young engineer at the institute whose brother had been diagnosed with a rare and fatal neurodegenerative disease. Even though Jamie was not a biologist himself, Ralph told me, he had hit on an intriguing idea for an experimental treatment. Now he was going to top-notch authorities on ALS, people at Harvard and the National Institutes of Health and other capitals of the biomedical empire, to build a consortium of people who might test his gene therapy and try to save his brother.

"It's a great idea," Ralph said, "and the clock is ticking, so there's urgency. His last day here is one week from now, on Friday the twelfth. He's moving back to Boston, where his brother is."

"How much time do they have?" I asked.

Ralph explained that with ALS the rate of decline is unpredictable. Jamie hoped to save his brother before he was paralyzed. "They may only have a year or so."

I do not remember what I said to that, but Ralph must have heard the doubt in my voice—so little time. He told me again that Jamie was doing amazing work for someone who had started his race with so little biology.

"He sounds like a prodigy," I said.

"I'll be very sorry to see him go," Ralph said. "Jamie's a delightful guy. He flies in twenty directions at once. He'll switch topics quickly. Your note-taking hand will get sore."

Ralph and I had spent many days talking about time, love, memory, DNA, the future of biology. He knew that when I am reporting, I scribble until my right hand can hardly hold a pen. Later on that year, when Stephen and I were riding around Newton in the cab of his pickup truck, with Stephen driving and talking and me scribbling notes, the weakness of my right thumb and forefinger would give me a small hint of what he must have felt when he stood outside his house with the key in the door.

"The whole family is colorful," Ralph went on, wooing me into the story. "The brother is a carpenter. . . ."

Ralph had no way to know how much the Heywoods' story was already stirring me. In those days, though my mother had been sick for some time, I rarely talked about her illness, and only with my closest friends. Even then I felt guilty, because my father was outraged and anguished that I would let our family secret that far out into the world. I knew that Ralph had to feel more than he was saying, too. His wife, Dani Grady, is a survivor of breast cancer and a national patient activist. So Ralph and Dani were bound to be moved by the Heywoods' crisis.

I hesitated. Then I asked Ralph, "What chance do they have?"

"Overall the chances are not great," he said, with a balance in his voice of pity and professional judgment. "But I don't think it's futile. Assuming it's not harmful to do this gene therapy, the idea could actually *work.* They won't know until they try it."

Ralph explained a few of the steps and hurdles that lay ahead for the Heywoods. Injecting DNA is a routine procedure in biology and biotechnology. A student in a fly lab let me try it when I was writing *Time, Love, Memory,* and it was as easy as playing a video game. He put a fly embryo in a video microscope so that I could see it in front of me on a big television screen. In the screen's lower right-hand corner I could see the tip of a very fine glass hypodermic needle. I had a joystick in my right hand, and when I jiggled the joystick, the needle moved around on the screen.

Slowly, I maneuvered the needle until it touched the embryo's rear end. Then I pressed a foot pedal. On the screen, I saw a shimmer above the needle tip, like hot air above a candle. And that was it. If any of that syringe-load of DNA found its way inside the right cells of that embryo—if a strand of DNA wriggled into one of the young cells that would become that fly's sperm or eggs—then the DNA I had just injected might pass into that fly's children. In principle, if the fly had a genetic disease, then in the fly's children the gene might correct the defect.

That kind of procedure is called germ-line gene therapy, because the DNA finds its way into germ cells—into sperm and eggs. With the same foot pedal, joystick, and microscopic needle, I could have performed germ-line gene therapy on a worm, a mouse, or a human being.

The procedure that Jamie was inventing for Stephen is called somatic gene therapy, from the Greek word for body, *soma.* In somatic gene therapy, the injected DNA finds its way into muscle cells or liver cells or nerve cells. But injecting DNA into just the right cells in a sick human body can be much more complicated than injecting an egg or an embryo. No one can do it alone. Jamie would need a big team of specialists.

"Putting together teams is Jamie's forte," Ralph said. "He'll be the catalyst. He'll approach people to make things happen. It's not easy to get scientists to line up like this. That's part of what's interesting. It's all quite unusual. Normally, ideas from outside are amorphous and rarely work. This idea is such a good one.

"A reporter could be helpful to the project."

I told Ralph that he might be showing me the future once again—or at least the next few years of my own. That feeling had been growing as we talked. Behind our science-fiction conversation about injections of DNA, I could hear classical music—all through the phone call, the rise and fall of passionate strings.

"Is that chamber music?" I asked at last.

"Yes, a Brahms trio. And fittingly enough for a phone call from the future, it's playing on my computer."

I read over my scribbled notes, and thought about Providence, which is only a short drive from Newtonville. Ralph had told me that the month before, Stephen had been diagnosed in Boston by Robert Brown of Harvard and Massachusetts General, a world authority on ALS. That same week, my mother had been examined by another neurologist there, Dennis Selkoe, a world authority on Alzheimer's and its relations, like Lewy body dementia.

Selkoe agreed with the neurologist my mother had already seen, Stephen Salloway, that she might have Lewy body dementia (LBD), and I had forced myself to read a few papers about it. Lewy bodies are tangles that clog neurons in the brain until the victim suffers falls and hallucinations. For some reason, people with LBD often see crowds of little elves or fairies who sit on the couch with them, or scurry around their ankles, run up and down the halls, dance on top of the couch cushions, and chatter to them. Patients do not seem particularly upset by these hallucinations; they do not even get upset if their families tell them there are no elves on the couch cushions. The English language actually has a word for what people with LBD believe they see: a smytrie, a collection of numerous small individuals.

I suspected that my mother's delusions were closer to garden-variety paranoia. When we were alone in the kitchen in Providence one day, I asked her if she ever heard or saw any little people around her. She said no. No smytries in her life. I was afraid my questions would make her angry, but she seemed almost flattered by my interest.

People with LBD can suffer from delusions other than smytries, and also from depression and a tendency to fall. Some of them also walk with an odd shuffling gait, as my mother began to do a little later on. All those symptoms did fit her condition. But as far as I could tell from the literature, doctors were not sure if those smytries were part of the same syndrome. It was shocking how little they did know. LBD was a fairly recent diagnosis. Although more and more doctors believed in it, LBD was not included in the International Statistical Classification of Diseases (ICD-10) or in the Diagnostic and Statistical Manual of Mental Disorders (DSM-IV).

In any case, my mother and Jamie's brother had been at Massachusetts General for their diagnoses at the same time. I still think about that: Stephen and Ponnie passing each other in the hall, a tall carpenter with a square jaw who looked as if he would always be young, and a small woman in a tailored blazer who looked as if she would never be old.

I was working with an editor at *The New Yorker* on an excerpt of *Time, Love, Memory.* He had asked me to propose another story for the magazine when we were done. I paced around my office. Then I sat down on the carpet with my back against a filing cabinet, rereading my notes from Ralph's call. Finally, I dialed the number that Ralph had given me: Jamie Heywood's office at the Neurosciences Institute.

The voice that answered the phone sounded very young to me, but husky and roughened by exhaustion. Jamie told me that he had just gotten back from another trip home to see his brother and to arrange for the move back east. "Typical pissing Boston weather," he said.

Jamie sounded nothing like the scientists I usually interviewed. He was all business and he talked at warp speed, just as Ralph had said. He told me he was maneuvering, wheeling and dealing, talking about patents and "heavy-duty intellectual property." He was meeting with ALS activists and with people in biotech companies like Cell Genesys, trying to get the ball rolling. His voice had a quality that I thought I recognized. It was a voice that sounded strong from having taken in the worst and still resolved to fight for the best. I heard what we are always trying to define: grace under pressure, action in the face of suffering, the will to hope in the teeth of despair, courage.

"The landscape is very complicated," Jamie said. "This is a forced journey into a new world. A lot of amazing people in it."

Even though we had not met, I found myself caring about him. "How are you holding up?"

"I'm in a slight depression right now. I'm trying to align people's interests in a way that won't hold up time—because the only thing that matters to me is time." Then I heard his spirits rebound—or rather, I listened as he talked his way back up, as if he could buoy himself by an act of will, or sell himself hope by selling it to me. There was theater in this: By showing us both how well he could sell hope, he was showing that he could sell others. It was that very power to sell hope that was the story. That theatrical power of Jamie's was now, in fact, Stephen's only hope.

"When I started, I had a million-to-one odds of getting something off the ground," Jamie said. "Then, OK, maybe ten-thousand-to-one odds. Now I'm maybe ten to one, or five to one. So I'm feeling pretty good. This is my kind of problem. I like this! Stops you from thinking about things."

I knew Jamie was not really five to one. "And how is your brother?"

"My brother is in an amazingly good state of mind. He was never a big planner. He takes things as they come. He flows well. He says, 'I had a great life, can't complain.' He used to paint. He wishes he could do that now, but he can't." Jamie told me that Stephen was having the most mundane and absurd problems getting out of bed, into his clothes, and through simple routines in the bathroom. Managing a zipper took ingenuity and improvisation. "That's what's on his mind more than anything else," Jamie said. His brother needed all his problem-solving skills just to get through the day.

"He can't decide if he wants to settle down or sow wild oats," Jamie told me. Stephen had found a girlfriend in Boston. Now he was wondering: Should he marry her and buy a house in Newtonville and make a baby? Or should he go out and find the biggest, baddest Harley there ever was, and roar off cross-country one last time? The Heywoods were a tight family, Jamie said, and he felt confident that his brother would decide to stay. "We're going to buy a house and renovate it. We're not good at sitting still.

"It's an extremely desperate patient population," Jamie went on. "There's nothing science is offering them but a twenty-four-month death sentence. No options, nothing out there. A ton of drugs, and none of them are worth dick." He told me that if I decided to follow his story, I could not write anything for a while. "It would be immoral to say anything until this has some hope of working." But one of the world's leading ALS researchers was already on board. (He meant Jeffrey Rothstein.) "We could be in business in a month. Evidence may exist in two months, or three months."

If I did follow the story, Jamie said, he would love to know what the doctors told me about him. "I don't know if they think I'm nuts," he said. "They're spending far more time with me than I would spend with myself. I can't conceive my role in all this, so I don't know how they see it. But I haven't found barriers." Every time he phoned an ALS researcher, Jamie said, he felt scared. "I get this burning, prepare-to-be-embarrassed feeling. It's *that* that's the barrier. People don't have the balls to ask the experts questions. But I'm learning."

Gene therapy, more than any other field of medicine, had the aura of the future about it, the ring of impossible promise, the feeling of the new millennium. On the other hand, I pointed out, it was an extremely controversial procedure. Jamie admitted that regulations could be a problem. "I've assumed that if I couldn't get approval, I'd go to Greece and pay someone there. Or Jamaica, or somewhere. But I think we'll be able to pull it off. It's doable," he said. "Maybe in a few months."

I asked Jamie if he had seen the movie *Lorenzo's Oil.* He had not. "I'm restricting the movies I see at the moment," he said. "No one's allowed to die anything other than a violent death. I haven't been sleeping at all for a month."

Jamie spoke of burning tension, nausea, rebellious bowels. Then he rebounded again and sounded strong and cheerful.

"One would not choose to get perspective this way," he said huskily. "I'm walking out of a hundred-thousand-a-year job now. I'm turning away far more. God, I feel so free it's astounding."

Jamie spent his last days at the institute cramming. He talked for hours about neurons with Joe Gally and about drug development with Ralph Greenspan. Within a pharmaceutical giant like Merck or Bristol-Myers Squibb, developing a new drug takes five or ten years. Economies of scale prevent the giants from focusing on ALS. With only 25,000 ALS patients in the United States, ALS is an orphan disease. The problem is common enough to serve thousands of academic biologists as their justification for basic research. When they apply for government research grants to study the workings of nerves and muscles, scientists point out that it might help lead to a cure for ALS. But ALS is not common enough for those researchers' findings to be seized upon by Big Pharma. And biotech companies, which exist to fill that gap, do not do enough to fill it. Translating research into treatment should be straightforward, like translating thought into action, but here the biomedical empire was inept, almost paralyzed.

These diseases really are orphans. The system lets them down. Jamie had long talks with Ralph's wife Dani, too. She told him that what is missing is not necessarily money for research, but money to move from research into development. Doctors treat patients with drugs and they discover ideas that could lead to new drugs; but they do not do the in-between work that could get their ideas to the market.

That is a gap that Jamie thought he could fill. He saw a niche open and waiting. Maybe his skills were needed here. He could not only cure

ALS, he could help to change the landscape. This is where the bridge needed building. He might even make a fortune while he saved his brother's life. He could imagine moving on from ALS to other orphan diseases, working as a catalyst to help cure disease after disease.

But to move into the open niche within a single year, he would have to solve problems on an extraordinary range of levels, from the molecular to the cellular and anatomical, and upward to the financial and the governmental. His parents were upper middle class, not rich, and he would need to raise a lot of money. He would need powerful collaborators. To inject a new drug into Stephen's body, he would also need approval from a series of regulatory agencies, all the way up to the Recombinant DNA Advisory Committee (RAC) and the Food and Drug Administration (FDA).

It is very difficult right now," Melinda wrote in her journal. "Things feel difficult. Jamie is preoccupied with how to go about implementing the ALS therapy, Stephen's therapy, his idea, money, lawyers on retainer, intellectual property, consensus, expenditures, money, researchers, publication, approval, animal trials, human trials. . . . I am worried because Jamie is worried. He brings it all home with him in his body, hunched up in his shoulders. I say it won't help to make decisions, no matter how difficult, with such a wound-up body. And last night I had brandy before going to dance, and Jamie had Chivas at Zorbas, and in general it's difficult, and he doesn't want to worry about me on top of everything else, and Dani said this kind of thing can break up relationships, and we disagree—it won't break ours, but it will splinter it a little, and small repairs will have to be made. I am going to Greece for three weeks. . . . I'm tired of my own worrying."

Sometimes in the middle of frantic packing, their minds would go blank. They would drop into what Melinda called "a white blind of mortality." One or the other of them would drift down onto their futon and curl into a fetal position. Melinda sliced a hunk of bread to eat with cheese and pomegranate jelly and the bread knife cut deep into the tip of her left index finger. Five stitches in the ER.

When Jamie and I talked on the phone, he kept our conversations off the record. He was not sure when, if ever, he could go on the record. And I was busy with book promotion and my parents' crises in Providence. So neither Jamie nor I was sure what would happen, but we agreed to keep in touch. Back when I was writing *Time, Love, Memory,* I had worried that readers would not get it: They would think it was a book about flies. Once at a cocktail party of scientists and science writers, I told Stephen Jay Gould what I was doing. He got the point before I had finished my first sentence. "Oh, that's a good way to do it," he said. But he also saw the problem. When he was writing *Wonderful Life,* his bestselling book about the fossils of the Burgess Shale, he worried all the time, he said. "Can I make people care about *invertebrates?*"

Well, I thought, this would be a story about vertebrates. It might be too sad to write, or to read. But here is a book about heroes with spines. I kept rereading an e-mail that Ralph sent me:

> Dear Jon,
>
> I hope it does work out for both Jamie and you to do the story. . . . I think that one of the things that makes Jamie's story unusual is the family's spirit of hope in the face of adversity and, in fact, with the odds overwhelmingly against them. The chances that their idea will work, if they have the opportunity to try it, are very, very slim. But what they are doing epitomizes the same spirit that Dani, my wife, proselytizes about to cancer patients: Be a thriver, not just a survivor; be an active participant in your treatment; investigate every possible avenue; push the medical system to its limit; and then, no matter what the outcome, you will not have been defeated and will have

lived life as fully (if not more so) than most. Jamie's and his brother's attitudes are in line with this and it's a very important message for the world.

Late that February, just before he left the Neurosciences Institute, Jamie and Melinda went over to Ralph's house and got drunk with Ralph, Dani, and his friend Tim Tully, a neuroscientist who had genetically engineered flies to have photographic memories. They all talked late into the night. Jamie was still not sleeping. By now he had come to believe that every day of his life was worth at least one month to Stephen, because he was convinced that within half a year at least thirty scientists would be working on his EAAT2 project and a few other ideas he was putting together. So for every day that Jamie lost, Stephen lost thirty days of laboratory research.

Ralph tried to comfort Jamie. *Don't think about that.*

But Tim Tully shouted him down. *No, he's exactly right! That's what it is!*

{ FIFTEEN }

Light Up the Cave

A few days later, Jamie and Melinda drove out of San Diego in a rented yellow twenty-four-foot truck. Jamie's hair was long and shaggy because he had not found time to get it cut. He had a nasty cold. His eyes and nose were red. He had a cold sore on his lip.

"Luckily, the truck couldn't speed," Jamie says. "You put your foot all the way down to the floor and went as long as you could."

Jamie's cellular phone service was giving him trouble, and each night at the next motel he was on the phone with AT&T. They stopped at a Best Western in Gila Bend, and a Motel 6 outside Oklahoma City, where he managed to hook up to e-mail. Melinda wrote in her journal, "Not being in touch with people on a daily basis may drive him out of his mind." She tried to keep up his spirits and hers. She wore glamour glasses and a French scarf and waved to other truckers from their cab.

Jamie says, "The country didn't feel that big, actually, which was surprising because I'd driven across before, and it felt very big before."

On March 4 the two of them and a few friends carried their boxes into John and Peggy's house on Mill Street in Newtonville through flurries of snow. That same afternoon, Jamie set up his computer and his phone among their half-opened boxes, using a crummy old door for a desk down in the basement, right outside the door of Stephen's Cave. Jamie began phoning doctors and neuroscientists in Boston, New York, Philadelphia, and Baltimore while Stephen watched, leaning against the plaster walls with his arms folded across his chest. Melinda noticed that he held his right arm behind his left, hiding and protecting his bad hand. He used his right hand much less often now, though he was still working on his parents' bathroom, grouting the tiles.

That night, Stephen gave Jamie two checks for five thousand dollars each, and told him to think of what he was doing as a business. He did not expect Jamie to cure him but he thought his brother might be able to slow down his disease—maybe preserve him in the condition of the astrophysicist Stephen Hawking. *If this works out and I have to be in a wheelchair for the rest of my life,* Stephen told his brother, *I'd like a little money.* He asked Jamie to use the cash to buy plane tickets, phones, whatever he needed to get into the business of curing ALS. *It's an investment.*

Because Jamie and Melinda had hardly begun to unpack, they did not know where to put the checks. Stephen propped them ceremonially on a shelf. He told Melinda that if she was willing, he would also pay her to do his books, and help him keep records of his daily medications. *My handwriting sucks*, he said.

Jamie felt naked placing calls down in the basement. "Before, I was the Director of Technology Development, calling from the Neurosciences Institute. And all of a sudden, I was just somebody." On March 8, to boost his confidence, Jamie got his hair cut at last ("a welcome event for all," says Melinda). On March 9, Jamie and his parents decided to start a foundation. They called it the ALS Therapy Development Foundation, a name that Jamie thought would convey weight and legitimacy to scientists. He felt as if he were still moving horribly slowly, with his foot flooring the gas. "The time-compression element here was unreal," Jamie remembers. "I would find myself saying, 'A few months ago we were at this point where we didn't know what to do,' and it was really *last week*. That's how strange time felt. Literally, we could go through lifetimes in two days."

On March 22, Jamie and Robert Bonazoli, the first of their friends to quit his job and join Jamie's crusade, took sledges to the basement walls. Jamie's father started down the steps when he heard the noise and they waved him back. *You don't want to watch.* They expanded the office and lugged in workbenches, installed outlets, extension cords, a printer, phone lines, a copy machine under the old map of Duck.

"That was the worst part for poor Stephen," Jamie told me. "Your privacy is already violated enough in this process. And then all of a sudden he would wake up in the morning and there would be two or three people working there right outside his bedroom."

Jamie began driving to laboratories in Boston, New York, Baltimore, courting people in competing laboratories, driving home during the night, and then spending twelve-hour days on the phone in the basement. The basement was "International Headquarters." Stephen thought Jamie had frightening amounts of energy. After a twenty-hour day, he was ready at 9:30 in the morning to do more work.

On March 25, Jamie made his first visit to Jeff Rothstein's lab at Johns Hopkins to talk about EAAT2. He drove home elated. Not only did Rothstein give Jamie several hours, but he walked him out of his office and down the hall to his next appointment. Jamie was thrilled, and he was delighted to have such good news. "That he would spend the time to do that!" Jamie told me. "He walked me all the way to the next lab. You go in and you don't know who you are or where you stand. And when someone makes a gesture like that—"

Another day a stranger called Jamie to ask him about his project. Jamie answered a few questions patiently but he had missed the name.

I'm sorry, who is this?

You want to talk to me.

It was Alexander d'Arbeloff, the chairman of the MIT Corporation, and founder and former chairman of Teradyne. D'Arbeloff is one of the most respected businessmen in Boston.

D'Arbeloff told Jamie to come by his office, and he gave him a few hours, too. He asked the kinds of questions that a tough, experienced manager asks when he is deciding whether to hire you or throw you out the window. Again and again he reminded Jamie that major drug companies take five or ten years to bring a drug to market. Jamie felt glum. But as he got up to go, d'Arbeloff said, *I'll give you a hundred thousand dollars.* It was the first of a series of long afternoons from which Jamie returned to the basement alternately uplifted and downcast. D'Arbeloff was even more intense and high-powered than Jamie, and he, too, did not have a doctorate. He was a self-made man. If Jamie had lost his sense of power when he quit the institute, he was regaining it now. "But this time it was internal," Jamie says.

Jamie went back to Grace Church. Stephen had always been a cheerful unbeliever and he was not about to change now, but Jamie began going with his parents every Sunday morning. He was comforted to see the bells of the church sitting out on display in the churchyard while the tower was being repaired. He had learned how to play those bells as a boy. Jamie told the new pastor that he remembered how he and his brothers used to run around in the great hall and slide on the floor.

"You'll be happy to know the children still do," the pastor said.

Again and again Jamie drove all day and came home late at night. He would walk through the backyard under the tall trees and in the back door. There were always projects going in the Heywoods' yard, and he might pick up an ax, a hammer, or a saw from the yard in the moonlight and lean it against the side of the back porch on his way in. Sometimes his mother and father waited up for him in the solarium, sleepy and warm, John in his pajamas, Peggy in her nightgown.

"We are lucky to have such a family, people tell me that all the time," Peggy says. "Well, we are lucky, but not as lucky as we could be, I suppose."

At this point I had not met Jamie and Stephen. I was about to meet Jamie; I would not meet Stephen until later. When I asked him about this time, Stephen told me that he could not remember how Jamie had broached the news about his project. "He started talking about it very hesitantly," Stephen said. " 'Well, what do you think about this? What would you think if I did that?' " The scale of Jamie's plans shocked Stephen, but he tried to keep his doubts to himself, and the more he listened the more he saw the point. The scientific knowledge to cure ALS might already exist, but it might not be applied in time to help him—not without an impassioned entrepreneur like Jamie.

When Jamie told Stephen about his gene therapy idea, Stephen wanted to say, *Well, someone would have thought of it already.* But Jamie explained that scientists were not rushing to try out their latest clues in the lab to save Stephen—even if it was an obvious idea, as straight-ahead and commonsensical as fixing a broken pump. "Jamie made it very clear that researchers don't work that way," Stephen told me. "There's such a lack of communication, it doesn't even matter if someone's already thought of it. It's irrelevant if *ten times* someone's already thought of Jamie's therapy. If nothing got put into motion, it doesn't exist."

I asked Stephen if he could tell me exactly how he had felt as he listened to Jamie describe his grand plan.

"It's not that I didn't believe it was possible," Stephen said. "Well, I guess I had doubts. Somehow I doubted that the idea for this particular therapy was possible. And so what if it was possible, how would it get done?" Jamie was turning their lives upside down for such a long shot. "Even if it might work," Stephen said, "I wasn't sure that's what I wanted him to do. What if I just wanted him to be in Boston, spend time with me, stuff like that?

"But it was very clear that it didn't matter whether or not he could do it, or whether I thought he should. I mean, he was so focused that you couldn't even have tried to stop him. It would've been counterproductive. Jamie could've easily gotten a job back here and stuff like that, but there's no way he would've. There was no stopping him. So that made things easier for me, I guess—easier than worrying about whether we were doing the right thing." When Stephen did bring up his doubts, Jamie gave him the same answer he gave their parents: *If I think there's even a possibility, I can't not do it.*

It was the fall of '99 when Stephen and I had that conversation. We were driving around Newton in his pickup truck. "I'm trying to think what on earth *I* did with the spring," Stephen went on. "I tell ya, I feel like it was a blur of working on the house a little and then hiding from the foundation, because it was in my basement. Oh my God, that's part of it, too. We redid the basement as well. Ugh. It was constant, and these guys are down there working like crazy. And Jamie's coming down at eleven o'clock at night to do shit, and I'm in there with Wendy, and I'm like, 'Just don't come down here right now.' And I'm spending nights at Wendy's, sleeping on her too-short bed and coming home with cricks in my neck.

"But that was fun, too, because it was all sort of new and exciting, buying cheap desks and a computer, and stuff like that."

Before Jamie arrived, Stephen had spent his time trying not to think about ALS. Of course, there were plenty of nuisances to remind him. He knew he was sick whenever he needed his right hand. "I guess I don't really ever think about dying," Stephen told me that fall. "I just think about what the future of the illness is like." Before Jamie moved down into the basement, Stephen had been able to look ahead very selectively, or as he put it, "in a very focused manner. I could think about it at certain times. The problem with the foundation stuff was that I lost that ability completely, because it was always being talked about. As Ben called it, 'The all-ALS channel, all the time.'

"What's so odd about this—" Stephen said, with a sardonic laugh, "is that ALS sort of speaks to my inherent laziness. I mean, it's a disease where you become gradually less and less able to get up and do stuff."

He was in love with Wendy, but everything was moving very fast—especially for a slacker. He still held on to his Cave, and she held on to her apartment. "Then Wendy came to me, and she had been nervous and stressed all day, and I said, 'What's wrong? What's on your mind?'

And she's like, 'It's stupid. I don't wanna tell you. It's dumb,' and I'm like, 'No. What is it?' for like four hours or whatever, trying to get it out of her, and she's like, 'Well, I'm worried about us having a kid if I'm not married, and I don't know if that's a good idea.' "

Stephen told Wendy he felt the same way. And so it was decided between them, more or less: marriage. "But it's odd," Stephen told me. "You know you want to have kids first. And then of course, wanting to get married because you want to have kids. I don't even know how to separate it out, and I wish I could, but there's nothing to judge it against. You can't say, 'Well, this is what would've happened, you know, if things were normal.' It also doesn't matter."

He no longer wanted to ride off on a Harley. Sometimes he still drove around Boston in his pickup truck looking at handyman specials. But on most days, while Wendy was at work, he sat in front of the computer and played games. "The computer lets me turn my brain off for a few hours," he would say. "Maybe it's a waste of time, but who gives a rat's ass."

One bright afternoon in March 1999, Jamie called me from his car. He had just driven down from Boston to Baltimore to see Jeff Rothstein at Johns Hopkins. Now he was driving back to Boston. My home, Doylestown, Pennsylvania, was not really on his way, but Jamie made it sound as if he would be right next door. As long as he was passing through, he wondered if we could meet in person.

I was sitting on the deck of a café in Doylestown when he came bounding up. Some people look marked for adventure. There is something clean and well made about them. They carry themselves as if they know how much they stand out from the rest of us, all humbly hodgepodge as we are. I could see at first glance that there was some calculation in his charm. But that came from desperation, I thought, and I was willing to be charmed.

We drank coffee and he told me his story as I have told it to this point. Even in that hopeless situation, he managed to carry himself like a winner. Every few minutes he excused himself—another message on his pager or his cell phone.

I knew it had to end badly. I think he knew it, too. But I still hoped that through sheer energy, will, and intelligence, Jamie would snatch some kind of victory in spite of everything—and not be destroyed with Stephen.

The longer I listened to him talk that afternoon, the more trouble I had keeping my head and remembering what Ralph had written about the slimness of Jamie's chances. And I still had some distance from the story. If I was having trouble, I wondered how Jamie could manage—young, speeding, unsleeping, with his impossible project to sell, and a brother to save.

{ SIXTEEN }

The Drowning Incident

Jamie is a light switch, on or off, and a few days later he closed down. An entrepreneur in California had given him a list of prominent gene therapists, and Jamie sat at his door-desk in the basement and stared at the list.

"He was stagnating," Melinda says. "Pulling himself like taffy." He paced outside the Cave melodramatically, crying, *What is the next step? What is the next step?*

He called a few of the gene therapists on his list, and they did not call back. A few others listened and pleaded full plates. Meanwhile most biologists told him that the field of gene therapy was hopeless.

"It is very easy when you are feeling insecure to be bounced up or down by how it goes," Jamie said, trying to explain those lost days and his frustration with himself. "See, I'd told Jeff Rothstein that I could get someone to make gene therapy. And then I had to go do it, and I was procrastinating because I was afraid to fail. There were a couple of discouraging conversations."

Gene therapy was already almost twenty years old. The field was begun in 1980 by Martin Cline, a specialist in blood diseases at the University of California, Los Angeles. Cline had thought of an experimental cure for thalassemia, an inherited form of anemia. He wanted to extract bone marrow from his patient, put it in a petri dish, inject a gene into the bone marrow, and then put the bone marrow back. Cline could not get permission at UCLA for that experiment, but he did it anyway with collaborators at clinics in Italy and Israel. So the field was born in scandal—and the gene therapy neither helped nor hurt the patient.

This had been typical of the field ever since: secrecy, controversy, a millennial aura, perpetual promise unfulfilled. In the United States, all experimental treatments are monitored by the FDA. But because gene therapy is so novel it is also monitored by the Recombinant DNA Advisory Committee (RAC)—an extra level of bureaucracy. About 4,000 Americans took part in about 400 gene therapy trials in the 1990s, mostly experimental treatments for terminal cancer. By the spring of 1999, nobody had been killed but nobody had been helped, either.

Small biotechs and big pharmaceutical companies were backing away from gene therapy. The Swiss drug company Novartis had put millions into a major gene therapy project for brain tumors. Novartis had just killed that project the year before. One of the big biotechs, Chiron, decided to stop developing new gene therapy projects in 1999. Many biotechs that had sponsored gene therapy trials in the 1990s had more or less folded by the spring of '99, and the stocks of the survivors were suffering.

It looked like a very hard year for the field. James Wilson of the University of Pennsylvania in Philadelphia was president of the American Society for Gene Therapy. Like everyone in the society he felt battered, and he was anxious about their latest round of experiments. So many years of hype, so little progress. The general feeling was that the year would make the field or break it. At the society's annual meeting in June, Wilson would tell the audience, describing one of the field's very few promising projects, a gene therapy for hemophilia, "The stakes are incredibly high." He was talking to his peers. He could be frank. "For once I may say what I really think: I hope to God this works."

Besides his list of gene therapists, Jamie had a list of scientists whose research might relate to his plan. One of them was a young biologist named Margaret Sutherland at Vanderbilt in Tennessee. She worked on epilepsy, not ALS. But she had genetically engineered mice that made too much EAAT2 protein, the pump that fails in people with ALS. Jamie wondered what would happen if those mice were crossed with ALS mice. Would their extra EAAT2 make them live longer? Would it cure them? It would be a quick and dirty test of his gene therapy for Stephen.

Jeff Rothstein was willing to try that experiment if Jamie could persuade Margaret Sutherland to send him some of her mice. But she was very new to her field, and her research was moving slowly. After a few brief conversations with Sutherland, Jamie let that lead go.

"Jamie out of his mind with obsession, pinched and grabbing hair, worried, working in franticness," Melinda wrote in her journal on April 12, 1999. "I keep teasing him but he is on edge, on an edge without much humor." She had just finished going through their paperwork and their bills. They owed $8,000 in taxes. At least, staying with John and Peggy, they did not have to pay rent. "Thank God we are living here."

That day Robert Bonazoli joined the foundation full-time. On April 14, 1999, the foundation incorporated in Massachusetts as a nonprofit organization. On paper, at least, Jamie's project was real.

"It's official," Melinda wrote in her journal. "I only hope he holds together."

Jamie was checking in with me now almost daily, and I was calling biologists I knew to ask their opinion of his project.

I knew a neuropathologist at the University of Southern California, Carol Miller, whose mother had died of ALS. Carol's husband is Seymour Benzer, the hero of *Time, Love, Memory.* She and Seymour had taken care of Carol's mother in their home, and Carol had since begun to study ALS. In her laboratory she had tissue samples of her mother's brain.

Carol found Jamie's story fascinating. She told me she was very excited about Jeff Rothstein's papers and the EAAT2 gene. "It's a really good idea," she said. "I think he's got a really interesting gene, a really interesting gene." Jamie Heywood might not save his brother, Carol said. "But this is the way advances are made."

On the other hand, most biologists told me that Jamie's idea was crazy. When I mentioned it to a famous molecular biologist at Rockefeller, he literally rolled his eyes. "Most people feel this is the time to figure out why gene therapy is not working," he said, "not a time to be trying it on human beings. Of course ALS is a terminal, awful, horrible disease. So when someone knows that they have it, they are willing to do anything. It's a little like prisoners on death row. The question is whether we're ready for those experiments."

While I was calling on biologists, I heard a rumor. Jacques Cohen had invited a group of infertility specialists on a cruise in the Caribbean. Cohen was organizing a retreat for his group and some other fertility doctors. They wanted to meet privately to think about the future—to speculate where their science was taking the human species. They were surrounded by such publicity now that they wanted to get far away from reporters and bioethicists to talk blue sky in safety.

I spent a whole day composing a letter to Cohen, reminding him of our dinner at Princeton in 1997 and asking if I could come along on the cruise as an observer. I thought he might agree if I promised not to write about what happened for a while. And I thought the cruise might be a happier story to write about than the Heywoods.

I called Cohen at St. Barnabas, and after many phone calls involving the two of us, his hospital's lawyers, my agent, and her lawyer, we settled on the terms of a confidentiality agreement. I promised to write nothing about what I heard on the boat for three years. After that, I could publish whatever I wanted, as long as I did not give away any trade secrets.

I can still see Cohen's final fax as it came scrolling out of my fax machine. According to the program for the meeting, he and his friends planned to spend their week's retreat not on a chartered boat, as I had imagined, but on a floating resort, the *Galaxy,* which is owned and operated by Celebrity Cruise Lines. First came the letterhead of Jacques's group, "Institute for Reproductive Medicine and Science of St. Barnabas. Gamete and Embryo Research Laboratory." Then came the title of the conference, "*Embryo Cruising in the Galaxy: The Near and Far Future of Assisted Human Reproduction.* Venue: Somewhere in the Galaxy."

One night, not long before the trip, Jamie Heywood stopped by our town in Pennsylvania. He was on his way home from another visit with Jeff Rothstein at Johns Hopkins. My wife and I took Jamie out to dinner at a local restaurant, Paganini. He was floundering, he told us. All that he had going for him was his hope that he could keep people like me lined up in a row and listening to him. He was bootstrapping from nothing but his own faith in his ability to sell, and his faith was failing. I could see that this was true, and at a certain point in the meal I began listening less to his words than to his desperation. He felt as if he were swimming out to sea, he was saying, and the tide was sweeping him farther and farther out, so that he was terrified that if he could not save his brother, he would drown, too.

I had been listening to Jamie talk that way for almost a month. It was hard to hear, and I was beginning to wonder if Jamie had a story at all. Suddenly I wondered if all he would ever have to sell was desperation. Even worse, I wondered if this was all he really wanted—to keep people like me listening. Writers meet sad people like that over the years. As Jamie talked, I had a vision of him drifting farther and farther out into some horrible Never-Neverland, toward which he was trying with all his might to draw me.

Suddenly I felt exhausted. I was falling out of the reporter's role. I was so overinvolved with Jamie's hopes and fears that I might have been drowning with him. Trying not to let him notice, I put down my notebook and pen, and picked up my fork.

Deb, of course, had no way to know what was going on inside me.

She was meeting Jamie for the first time. All she saw was a handsome wheeler-dealer, and she hoped I would not get involved with him. Still, she heard Jamie speaking with passion and saw me eating my pasta.

"Shouldn't you be taking notes?" she asked.

Jamie's eyes and mine met for a moment across the table. I looked down again, but I was sure that he had seen what was going on with me. My face felt hot.

When Jamie stood up from the table and said good-bye before the long drive home—he still had another six hours to Newton—he told me that he would not call again until he had some progress to report.

"Something will happen," Jamie promised me with a level look, eye to eye. He was speaking directly to the unspoken thoughts he had seen in my face, and without the slightest defensiveness that I could see.

He said, "Nobody who knows my family would doubt that something is going to happen."

{ SEVENTEEN }

Somewhere in the Galaxy

"Veeee-tro!"

Jacques Cohen stood by the piano in the Stratosphere Lounge of the *Galaxy* with a tiny digital camera, making a big monkey-face grin. His in-vitro fertilization team and his guests stood together in rows on the carpeted steps of the lounge and grinned back. The camera flashed. And so began our cruise into the future.

The *Galaxy* was docked in San Juan, preparing for departure. Leaning on the railing of one of the upper decks, Cohen apologized to me urbanely for all his precautions and preliminaries, and the need for a confidentiality agreement. He explained that many of the people on his team had been reluctant to have a writer on board. Their guests from other infertility clinics were even more reluctant. The whole point of the meeting was to talk about the future freely, without having to worry about the press.

They had decided to invite Lee Silver along, but that was an easier decision. Two years had passed since the night we had met at Princeton and Silver had torn up the manuscript of *Remaking Eden.* Since then, Dolly really had changed Silver's life. He had quit molecular biology and moved across Princeton's campus from the Lewis Thomas Laboratory to the Woodrow Wilson School of Public Policy. He was lecturing in dozens of countries. He had spoken at the Vatican, under tight security. He was a frequent guest on *Frontline.* He was getting death threats. He was now one of the world's most public advocates for the kind of work that Jacques Cohen and his colleagues did in private.

Silver was at war with bioethicists. He thought they had no understanding of the real world and market forces. That spring a Canadian bioethicist, Margaret Somerville, made a speech urging this work to slow down. Silver sent an angry letter to the editor of *Nature* titled "Bioethicists Must Come Down to Earth." He wrote, "Any regular reader of your journal is sure to wonder on what planet Somerville has grown up. It is certainly not one on which science or private industry exist, for if it was she would surely know the lunacy of her proposition."

The *Galaxy* has a hugely jutting prow like a beak, which gives the ship the look of a showy yacht enlarged five hundred times. She is a Love Boat of 76,522 tons, 866 feet long, with four diesel engines, four auxiliary engines, bow and stern thrusters, fourteen decks, ten elevators, and a guest capacity of 2,217. In a ship that size, the conference of the Institute for Reproductive Medicine was inconspicuous.

The next morning in the *Galaxy* auditorium there were spiral notebooks and pens stacked neatly beside the gleaming regular and decaf coffee urns. There were heavy curtains drawn across the portholes, to allow for showing slides from the PowerPoint projector that Cohen had carried aboard with his luggage. Most of the talks were technical ("Cytogenetic Diagnosis: What do we do now?") and concerned matters like the latest techniques for the biopsy of embryos, the freezing and thawing of embryos, the implantation of embryos. Lee Silver sat just behind the PowerPoint projector with a new Mac laptop open on his chair arm, speed-typing his notes. "You realize that if any right-to-lifers were here," he whispered to me, "they would be going absolutely bonkers. What they're doing is experimenting on living embryos. This is illegal in most countries." It is also forbidden in the United States if the work is done with government grants from the National Institutes of Health or the National Science Foundation.

But because it was funded by Cohen's private clients at the IVF clinic, it was legal, and it furthered the cause that he returned to again and again: his goal of 100 percent implantation rates.

To me, the talks began to get interesting when Cohen's chief scientist turned to the question of longevity. Suppose the telomeres, the very tips of our chromosomes, really do turn out to be the key to long life, as some biologists have suggested. If that happens, should they begin selecting human embryos for the ones with the longest telomeres, selecting for Methuselahs?

"Aging," Jacques said musingly from the podium, after his chief scientist's talk, gesturing with his cup and saucer at a slide on the wall. "Aging is a *disease*—but not considered as such by ethicists. If you're a potential parent, would you not be happy to give the best to your child?

"*Do* we do this?" he asked. In his clinic they already selected babies based on their odds of living an average life span. Should they begin to select babies based on whether they would live longer than average? "And do we talk about this in public?" This would be the first time their clinic moved from selecting embryos to avoid diseases to selecting embryos to enhance their quality of life—unless the world did agree to call aging a disease. "The question is, how to proceed with this in terms of advertising so we don't get shut down."

"You have to realize how fast people go through enormous opinion changes," Silver said reassuringly from his seat by the projector.

"Do we want to do this?" Cohen repeated. "I mean, we all know who will come first," he said, and gave the group a big, warm, amused, confidential grin. He was thinking of those die-hard clients, the rich and crazy ones, God bless them, who always want the latest.

That was Sunday. When I found my way to the lecture hall on Monday morning, I saw that Jacques Cohen was on the podium again, listening to a speaker. The lecture in progress sounded technical. I sat down in the back row of seats and tried to follow what was going on.

Speaker: "Mrs. Smith had genetic abnormalities. I can mention that name because it's so generic. Her name really was Mrs. Smith. . . ."

Cohen, a few minutes later: "And she chose to deliver just then, yes?"

Voice from audience, with satisfaction, as if to convey an arrival after a long haul. A foreign-accented drawl: "Ma-a-a-y-y-y 28."

They were talking about something called cytoplasmic transfer by injection.

I started listening more closely.

Cohen said he knew of only a few other groups who were *succeeding* in their cytoplasmic transfers. "The groups in California *announced* doing it—for PR reasons," he said. Then, in reply to a question from the audience, he said, "The parents are informed, they get many hours of training, they sign forms of very well-informed consent."

I strolled up to the PowerPoint projector and leaned over Lee Silver, who was typing madly on his laptop. "What's going on?" I whispered.

Silver whispered back, electrifyingly, "*Gene therapy!* He's talking about the first babies born with genetic engineering. They make it *work.* I don't know if even they themselves appreciate that!"

This was the same project that Cohen had mentioned over our dinner at Princeton, when the world was talking about the birth of Dolly. Dolly was created by nuclear transfer. In that procedure, the nucleus is extracted from an egg, and another nucleus is transferred in. But these babies were made by cytoplasmic transfer. In this procedure, cytoplasm is extracted from a fertile egg and transferred into an infertile one. It is something like taking egg white from one egg with a syringe and injecting it into another.

Most of the DNA in a cell is packed inside the nucleus. But there is some DNA in the cytoplasm, too. The mitochondria, the energy packs that float in the cytoplasm, carry their own DNA. So the babies that Cohen's group made by cytoplasmic transfer had inherited DNA from two eggs and one sperm. That is, they carried genes from three parents.

So this was yet another kind of genetic engineering, as Silver said. You could argue that these babies represented not one but two firsts: the first successes of gene therapy, and the first germ-line gene therapy ever done on human beings.

After Cohen, the next speaker was Rick Adler, the team's microscopist. He explained that he had been examining the eggs in their cytoplasmic transfer project with the new wonder tool in the lab, the confocal microscope. Twenty-five years before, the wonder tool had been the electron microscope. But with that, you could only look at a dead, stained cell. With the confocal microscope, you could look at a live cell, and because the imaging is done by a computer, you could see things right away.

As Rick explained all this, there was a little *hmmm* of interested approval. "Turn the lights down a little bit more."

When Rick looked at the embryos in their cytoplasmic transfer project, he said, he noticed that the blebs, the fragments around each embryo, were extremely "hot." "Lit in red, in this picture. Lots of mitochondria there. This is the embryo I happened to be looking at when Cohen walked by the room," Rick said. "And he went *bananas.*"

During a break, I went out with Silver to an upper deck of the *Galaxy*, aft, and in the Orion Lounge, Silver told me why Jacques Cohen had gone wild.

"The question is, what allowed these women to get pregnant," Silver explained. Through the injection of cytoplasm, only a small percentage of the mitochondria entering the fertilized egg came from a third party. But in the embryo that Rick had examined, about *fifty* percent of the mitochondria had come from a third party. That was why the confocal microscope image had shocked Cohen. "The only way that could happen is they're being *selected,*" Silver said. The mitochondria from the donor mother had been selected in the egg as the embryo grew, while the mitochondria from the birth mother had been rejected. In other words, it was the mitochondrial genes of the third party that allowed that baby to be born.

"So it looks like gene therapy!" Silver cried. "They're scared to death, because it's genetic engineering and Jacques knows it. And the whole world thinks that's horrible, horrible, horrible. And maybe it's *not* so horrible, horrible, horrible. And that's maybe one reason Jacques wanted *you* here—to say that maybe it's not horrible."

During the next break, I found Cohen and Silver and a few others standing by the door to the auditorium. "This is genetic engineering!" Silver was saying, in a voice like a headline.

"We keep trying to avoid the term," said one of the young scientists in Cohen's group, Jason Barritt.

"Oh, it's genetic engineering. That's correct," said Cohen.

"Congratulations!" Silver cried, again in sixty-point type, and shook Cohen's hand.

"You have changed the genetics," Cohen continued, cautiously. "But not the *nuclear* genetics. Which is what everyone is worrying about. These are very *subtle* changes."

Cohen was reminding Silver that he had injected only cytoplasm, not a nucleus, into the egg. The third-party DNA came from the mitochondria, not the nucleus. The distinction mattered very much to Cohen. If the procedure they were performing was germ-line gene therapy, then it was banned by international agreements. But since Cohen had injected cytoplasm that contained mitochondria that in turn contained DNA, he felt that he had not broken the ban.

Silver himself seemed to think that was just quibbling. He was not interested in arguing whether or not Jacques had broken a ban. He was interested in the next step. He said he could imagine that giving the egg an injection of mitochondrial DNA from a healthy young woman could produce a baby with more energy, better endurance. He urged Cohen to select for those qualities in the cytoplasm donors.

Cohen was surprised. "So you think that we should actually *select* for . . . ?"

"If you need an egg donation," said Silver, "you might as well choose the best you can."

So now we were moving in one easy step from germ-line gene therapy as a medical treatment to germ-line gene therapy as an enhancement: the deliberate attempt to engineer better and fitter human beings.

Young Jason Barritt noticed me scribbling away, leaning against the wall. "He's writing his next book!" he warned Cohen.

"I'll choose stupidity," Cohen said to me, lightly. "Can you make a note of that? Ignorance is bliss."

"Your next book," Jason said to me again, with a significant glance at Cohen.

"That's OK," Cohen assured him. "He's under contract."

"There's an invisible zipper right here," I said, idiotically, running a finger across my lips.

"It's the *hand,*" Jason said. "I want the hand to slow down." He reached over and tapped my hand as I scribbled in my notebook. He was worried about the media, he told me. "No offense," he said.

"He's not the media," said Silver.

Cohen said he thought bioethicists' resistance to their work was probably going to get worse, not better. "I went to the Internet again to get slides for this week's talks," he said. "The anger is amazing. And these people are—" A small smile and nod to Silver.

"—our colleagues," Silver said.

They talked for a while about a few of the famous biologists who hated Assisted Reproductive Technology, ART.

"This started a long time ago," Silver said. "People rejected *medicine.* People rejected *vaccines!*" he said, speaking in the tones of one of Galileo's comforters.

"But we don't really want to be the ones rejected," said young Jason.

That night at dinner, Cohen, Silver, and most of Cohen's team and guests sat at Table 600 in the corner of the Orion Restaurant wearing dark suits and evening dresses. Cohen had been invited to join the captain's table for dinner, but he had chosen to dine with his team.

Sitting at the center of the table, with Silver at his side, Jacques Cohen raised his wineglass.

"To you," he said to Silver. "Thank you for joining our group this week."

Everyone drank the toast. Then Silver raised his glass.

"To more playing God!"

{ E I G H T E E N }

"My Husband, Doctor Frankenstein"

I had thought I was getting away from gene therapy. Instead I had sailed right into it. If that operation of Cohen's was the first genetic engineering of human babies, then it would have to be regarded as one of the most extreme tests of our bioethics. And when I went back to the Heywoods, I would see how complicated the ethics get even when we are talking about what might look like a very simple story at the edge of medicine.

But while we are on the boat, and talking blue sky, I have to pause to tell a story that I think represents one of the greatest extremes of all.

Genetic engineers like to say—I heard them say it several times on the *Galaxy*—that playing God with the fate of the human species is impossible. Genetic engineering is too expensive and labor intensive. There are billions of people on the planet. How could anyone ever engineer more than a small handful of fertilized human eggs? A dozen, a hundred, or even a thousand would be a drop in the gene pool. The drop would vanish without a trace.

I first heard that argument made by Lee Silver in one of his classes at Princeton in the Lewis Thomas Laboratory, a few years before the birth of Dolly. I was sitting in the back of the hall, as usual. The average age in the room was twenty-one, Mary Shelley's age when she published *Frankenstein*.

At that time, I was fresh from my time in the Galápagos. Sitting in the back of the lecture hall that night, I realized that the conventional wisdom underestimates our power to influence human evolution.

In the Galápagos, it did not take many finches to start a new line: All it took was one lost flock. A million years ago, a few birds were blown out to sea on freak winds, and landed on the bare new lava of the islands. There they were cut off from their kind. At first they were isolated only by geography. Then they were isolated by their genes. They evolved what is known in the jargon as reproductive isolation. They diverged, and diverged some more, and today there are a dozen species of Darwin's finches.

To create a new human species, genetic engineers would not have to engineer billions of babies. All they would need is to engineer reproductive isolation in a few babies. All they would need is fifty Adams and fifty Eves. No one could predict the fate of those hundred babies, of course, but by definition they would be a new species, as separate from the rest of us as Neanderthals, *Homo habilis,* and Cro-Magnon.

A few days after I had that notion, I dropped by Silver's office in Lewis Thomas and asked him what he thought of my idea. He listened for a few moments and said, softly and dramatically, "Shut the door."

"It wouldn't be easy, but in theory it could be done," he told me. "It's a fantastic idea." And on the spot, he thought of a way to do it. "There's a case in mice," he said. In 1972, in a certain lonely valley in Switzerland, the Valle di Poschiavo, Silver said, a biologist discovered mice that have only twenty-six chromosomes. Normal mice have forty chromosomes. The mice in that valley look exactly like normal mice, and they have exactly the same genes as normal mice, but a few of their chromosomes have shuffled and fused. They have the same encyclopedia, so to speak, with a few volumes mis-shelved or stuck together. "But the incredible thing is that these are new mice," Silver said. "They can't breed with other mice." In the jargon of evolutionary biology, they are a cryptic sibling species—a hidden new line of life. They have evolved reproductive isolation.

Lee thought that in principle, genetic engineers could do something like that to a very young human embryo, say a male. They could go in at the eight-cell stage and make pieces of chromosome 2 trade places with pieces of chromosome 13. They could shuffle three or four volumes of the encyclopedia. Then they could engineer the identical shuffle in the chromosomes of forty-nine more male embryos and fifty female embryos. When they grew up, those fifty Adams and Eves would be able to make babies only with each other. Their line would be cut off from the rest of the species, just like the mice in the lonely Swiss valley. Misplaced chromosome segments like these are known as chromosomal translocations. Evolutionists have studied translocations for years as one plausible mechanism in the creation of reproductive isolation and the origin of new species. Once reproductive isolation occurs, the myriad random mutations that take place in each generation can add up, and the isolated population and the main population slowly diverge.

"That's the way I would do it," Silver said. "No reasonable person could be in favor of this happening. But biologically it could be done."

I thought that over. Then I called a few other molecular biologists. Each one thought it was a wild idea and on the spot each one thought of another way to create an isolated human species. Molecular biology was moving so fast that it was capable of experiments as outrageous as anything in the literature of science fiction. Naturally, I thought of that dark night in another valley in Switzerland, in 1816, when Mary Shelley laid her head on a pillow and dreamed her famous nightmare. She saw a young man, she wrote afterward, "a pale student of unhallowed arts," kneeling beside a body that he had put together in secret. The body stirred, and the student ran off, hoping that the thing he had made would die again, praying that while he slept the strange spark he had lit would go out. "He sleeps; but he is awakened; he opens his eyes; behold, the horrid thing stands at his bedside, opening his curtains and looking on him with yellow, watery, but speculative eyes."

I also thought of the celebrated dark night in England in 1885, when Robert Louis Stevenson thrashed in his bed and cried out in horror. Frightened as he was, when his wife woke him up he was bitter and indignant. "I was dreaming of a fine bogey tale!" he said. In his dream, a doctor had mixed a potion and transformed himself into a murderer. That nightmare was the beginning of *Dr. Jekyll and Mr. Hyde.*

No one could create a new species of human being in 1816, or change human nature in 1885. But now pale students had new tools. Early in 1995, the *New York Times* had run a story by William Stevens with the headline: "Evolution of Humans May at Last Be Faltering." "Is human evolution ending," the story asked, ". . . Or will humankind, armed with the tools of molecular biology, seize control of its own evolution?" I thought I knew the answer.

Mary Shelley did not know anything about chromosome translocations, of course. She dreamed her dream several decades before the monk Gregor Mendel inferred the existence of genes in his monastery garden. So she built an adult from spare parts, rather than an egg from spare parts. Otherwise the two scenarios are essentially the same. In the novel, Doctor Frankenstein's creature begs him to make him a bride: "My companion must be of the same species and have the same defects. This being you must create."

So much for the pieties of the One Hundredth Psalm: "Know ye that the Lord he is God: It is he that hath made us, and not we ourselves."

I felt a creator's pride in my idea. I was experiencing firsthand the glamour that even a very ugly thought can acquire when that thought is your own. For a few days, I understood what the atomic scientist meant at Los Alamos when he spoke of the "technically sweet." I sat up late into the night scribbling until the nearest *Homo sapiens sapiens* rolled over and murmured, "My husband, Doctor Frankenstein."

A few days later I thought I found a fatal flaw in the plan. When I got out a few genetics textbooks and looked up chromosome translocations, I saw that moving chromosomes around is dangerous. Some people who make babies with Down's syndrome have a piece of chromosome 14 on chromosome 21. People who die of Burkitt's lymphoma have a piece of chromosome 8 on chromosome 14. People with still another misfiled fragment are born with a syndrome called *cri du chat,* the cry of the cat. They have tiny heads and thin, high-pitched wails, and they die in infancy. If you chopped up human chromosomes and moved them around, you might hurt the babies, and you might not know the damage until they grew up.

Suppose you changed only a single gene in the sperm and a single gene in the egg to alter one of the chemical signals by which they recognize each other. Then the sperm of your Adams could meet only with the eggs of your Eves. But you might not know what other kinds of work that particular signal does in the human embryo or in the adult brain. You could not know with certainty what your little project might destroy in your progeny, until you tried the experiment. Who knows? You might even alter the moral sense, as Dr. Jekyll altered Mr. Hyde's. You might change something that is deeply and subtly human, so that the adult who sprang from that altered egg would give everyone near him a sense of obscure inner deformity, as Hyde repelled everyone he passed in the fogbound streets of London.

I stopped by Silver's office to tell him that the Adam and Eve project could not work—but he dismissed my objections. He said one did not need to understand the evolution of reproductive isolation in order to engineer it, any more than one needed to understand the evolution of flight in order to build an airplane, or the evolution of the brain in order to build a computer. He thought it would be straightforward to avoid harmful translocations, for instance, and to make only benign ones. "Millions of translocations and rearrangements happen all the time," Silver said. "Most of them do absolutely nothing." In fact, he said, in many infertility clinics, men and women are checked for these chromosome translocations as a matter of routine.

He pulled down a book from his shelf, *Infertility in the Male,* by Anne M. Jequier, and opened it to chapter nine. According to the book, men walking into infertility clinics are eight times more likely to have translocations than men who are fertile. In most cases, there is nothing else wrong with them, nothing at all. They are as healthy as the mice in that Swiss valley. As I stared at the page of statistics, it dawned on me that there would be virtually no guesswork involved in choosing translocations for the Adams and Eves.

"You could copy the translocations from the records at an infertility clinic," I said.

"You're right! I didn't think of that!" Silver said. "You actually have the experiment in hand."

We both laughed tight little laughs.

"This is scary," Silver said. "This is very scary." In a way, he said, it reminded him of 1975, when molecular biologists were so unnerved by genetic engineering that they called a moratorium. "They thought they were playing God *back then!*" Silver said. "But now we've reached another point. They were fooling around with human genes. They weren't fooling around with human beings."

The creation of a new human species is one of the many strange possibilities that are now in our hands. It could be done. In the end, the only ineradicable problem in the project may be the one that Shelley and Stevenson both foresaw from the beginning, the flaw that gives their novels power: the pathos of the creatures themselves. No matter how well you made your fifty Adams and Eves, they could never mate with anyone but each other—breeding in a bottle, like fruit flies, with no escape. They would have every right to be furious. They might gnash their teeth at their creators. They might cry out as Adam does in *Paradise Lost,* in the lines that Mary Shelley chose as the epigraph to Frankenstein:

> *Did I request thee, Maker, from my clay*
> *To mould me Man, did I solicit thee*
> *From darkness to promote me?*

Part Three

The Construct

Who is strong enough to rule the sum of the immeasurable; who to hold in hand and control the mighty bridle of the unfathomable?

Lucretius

{ N I N E T E E N }

The Cowboy

I thought I would leave you alone until I got something done."

It was Jamie Heywood. He was calling from a cell phone in his car. "We're feeling good," he said. "Things are going."

Jamie sounded level and confident. When I heard how much he and his family had done since we last spoke, I was amazed.

It was summer now, the summer of '99, and from what Jamie told me, the Heywoods were all having a hectic ride. I had to watch. I pitched the story to my editor at *The New Yorker,* and went to work.

Just after our dinner at Paganini in April, Jamie had been sitting at his door-desk in the basement with Melinda and their friend Robert Bonazoli when the phone rang. It was one of the gene therapists on his long list, returning his call. The name of the scientist, Matthew During, meant nothing to Robert and Melinda, but Jamie knew that he was one of the boldest and most controversial young doctors in the field. He was also the one whose work came closest to the genetic engineering project that Jamie had invented for Stephen.

Matt During had been trying for a few years to develop gene therapies for neurodegenerative diseases, including Alzheimer's and Parkinson's. The project that had made him famous, or notorious, had begun in 1995 when he was a young, ambitious doctor at Yale, still in his thirties. He was working there with another neuroscientist, Paola Leone, and she was young and ambitious, too. The parents of a baby with a rare and grotesque neurodegenerative disease, Canavan's, had approached the two of them and persuaded them that Canavan's would make an ideal disease on which to try out and refine the tools of gene therapy. If During and Leone could figure out how to get the right piece of DNA into the nerve cells of the dying baby, they would know right away by certain simple diagnostic signs whether the DNA was working—even if they did not save the baby. Then they could do the same thing with different DNA injections for patients dying of many of the other more common horrors like Alzheimer's and Parkinson's. They could use the Canavan baby for a Phase 1 safety trial: a first step toward a treatment. But of course the baby's parents, a doctor and a psychologist in New Fairfield, Connecticut, hoped that by some miracle the experimental treatment would do more than help these young doctors refine their tools.

Matt During and Paola Leone had thrown themselves into the design of a gene therapy for Canavan's. The clock was ticking because a Canavan baby is born healthy and then becomes rapidly paralyzed, blind, and retarded, and begins to suffer from seizures. Such babies die by the time they are five, ten at most, but they are gone long before that. During directed the science on the team and Leone was one of his postdocs. Their project would have to get the approval of review boards at Yale and then in Washington. In Washington they would have to get their proposal past the RAC, and then the FDA. But if they were to help that baby, they had no time.

During and Leone believe that gene therapy will cure these incurable diseases and many others. They dream of revolutionizing the practice of medicine. Like Jamie, they have the outlook of outsiders, rebels with a cause, partly because they are young and ambitious and partly because they each come from islands far from the centers of the biomedical empire: During from New Zealand and Leone from the island of Sardinia. During was being courted at the time by the

University of Auckland in New Zealand, and he decided to leave Yale and carry out the Canavan experiment there.

At Auckland, he injected DNA into two Canavan babies from America, one of them the baby from Connecticut. By now, parents of many other Canavan babies in America were besieging the young doctors to inject DNA into their babies. So even before the first round of gene injections, During and Leone had begun planning a second round, this time for eight American children with Canavan. This round, like the first, was designed to test whether the procedure was safe.

The work made Matt During instantly famous in New Zealand as the first pioneering geneticist there to do gene therapy. Back in the United States, a member of the RAC happened to read about the trials in a newspaper, and decided that During and Leone had gone around Yale's ethical review board, and the RAC, and the FDA, to treat these American babies in Auckland. A story about this accusation appeared in the British medical journal *The Lancet.* During protested. He said he had gotten approval for his plan from authorities at Yale and in New Zealand. But *The Lancet* quoted someone on the RAC as saying that his claim was "inaccurate to put it politely." An ethicist at Otago University in New Zealand charged that since the babies could not give consent and since the trial could not cure them but was purely for the future, During and Leone were "using children in a manner which is not acceptable under any international criteria."

It became one of the celebrated cases at the edge of medicine. Matt During was forced to delay the second round of injections. He went back to Yale, and parents in America found senators to pressure the university to approve the experiment. The board of Yale's biosafety committee met again and again, and could not agree to allow it. The babies were doomed and their parents were desperate; but gene therapy had never saved one human being. An article in *Nature* that fall reviewed all of the gene therapies that had ever been tried. That year, doctors were injecting their patients with genes to treat or cure cancer, heart disease, and dozens of rare genetic diseases. "Although more than 200 clinical trials are currently under way worldwide with hundreds of patients enrolled, there is still no single outcome that we can point to as a success."

Which was true, unless you counted the babies that Jacques Cohen had brought into the world by cytoplasmic injection. But gene therapists did not know about that.

Eventually the Canavan controversy made both *60 Minutes* and the cover of the *New York Times Magazine.* The article in the *Sunday Times* was written by the reporter Michael Winerip. He said Paola had shouting matches with one of their older, more conservative colleagues at Yale, a doctor with the lovely name of Margretta Seashore. The magazine's cover was designed to grab and shock. It showed a snapshot of one of the Canavan babies whose parents wanted him in the second round of injections. He was a toddler named Jacob Sontag, a very little boy with a face pathetically thin for his oversize, thick glasses. The text read:

> Their 19-month-old boy was deteriorating, and Richard and Jordana Sontag had come to Yale, desperate and angry. An experimental gene therapy was their only hope, but the scientists were not sure it was safe. The couple spotted Dr. Margretta Seashore in the lobby. "Look at him," Richard said, thrusting Jacob in her face. "You tell him why this protocol is delayed while he's dying. . . . You look at my son dying. Tell it to him."

Pressure built at Yale. Jacob Sontag received the injections of DNA on January 22, 1998, and a second round of injections that September. By now the boy was two and a half. He and the other babies who got the injections continued to decline, although some of the parents believed they saw a few subtle improvements. So did Paola, a passionate scientist who often fell in love with Canavan babies and their families. A paper that Matt and Paola published later on in *Annals of Neurology* presents evidence that their gene therapy had made the slight but measurable differences in brain chemistry for which they had hoped.

The two young scientists left Yale. They were given a large laboratory and a generous welcome at the Jefferson Medical College in Philadelphia, which is a strong medical school but one tier down from Yale. To attract them, the school allowed Matt to set up his own center there, the CNS Gene Therapy Center (CNS for central nervous system). The school allowed him to keep his appointment in Auckland, where he still ran a large laboratory.

Jamie Heywood knew all of this, having done his homework. He was electrified when he heard Matt During's name on the phone. So far, of course, Matt and Paola had saved no one with gene therapy. Still, they had carried out the world's first gene therapy for a neurodegenerative disease of the brain. And best of all, from Jamie's point of view, they were fighters and fast workers. From the day that Matt and Paola were approached by the first parents of a baby with Canavan's to the day they were injecting DNA into that baby's brain was only twelve months.

Down in the basement, when Jamie took the phone to speak with During, Melinda and Robert could see that he was *on.* He was projecting his confident voice, highly caffeinated and candidly seductive at the same time. And Jamie did feel very sure. He was ready. After pitching his gene therapy idea to more than sixty people, he had learned the lingo. He knew enough to engage Matt During properly.

When he hung up the phone, he looked at Robert and Melinda.

He said, *We've got the guy.*

On April 23, 1999, Jamie met Matt During for lunch at Jefferson in Philadelphia. From their phone talk, Matt had gotten the impression that Jamie was a biologist. Even so, the two of them sized each other up and clicked immediately. They are both pragmatic, urgent men with adventurous grins. In fact, they look like brothers: boyish and very fit, with short-cropped brown hair and sharp cleft chins. Matt was forty-two that year, but he looked only slightly older than Jamie. Like the three Heywood boys, he is an athletic dreamer with an air of slightly raffish competence, a sort of impatient willingness to dare.

Jamie, with his strategy of WIIF'M, had decided exactly what assets he brought to Matt During. He had his connections at the Neurosciences Institute and even stronger connections at MIT, through his father and Alexander d'Arbeloff, chairman of the board of trustees of MIT. Matt had studied at Harvard and MIT and worked at Yale; anyone who met him at that time could see right away that he was feeling slightly defensive about his new berth at Jefferson Medical College. Jamie also had his connection with Bob Brown at Harvard, and the enthusiasm of Jeff Rothstein at Johns Hopkins, who had promised to help if Jamie could find a gene therapist. Besides these assets, Jamie could offer his own skills as a project manager and entrepreneur. He could also offer a patient who would risk a dangerous treatment if he gave the word.

An explorer is always looking out for ships of opportunity, and this looked like a good ship to Matt During, who was used to talking with families of desperate people. He could see that Jamie had studied the disease and the playing field, and he liked the offer. Over their lunch he told Jamie that he was interested in exploring Jamie's idea for a gene therapy, and if they decided to proceed together, he had the resources to help. In Auckland, where he spent part of every year, he had sixteen or seventeen people in his lab; and that year he had fifteen at Jefferson, along with three PhD students from Auckland who worked with him there. All this was true, although Jamie could see that Matt's center at Jefferson was still new and raw, a meandering and disheveled maze of half-furnished rooms and crowded hallways lined with opened and unopened crates and stacked boxes of surgical latex gloves, even a lost-looking couch. The place felt as if the movers had just trooped through and were about to come back.

With their colleagues in Philadelphia and Auckland, Matt and Paola were now developing gene therapies for Canavan's, Parkinson's, diabetes, epilepsy, Huntington's. Matt was preparing a paper about the improvement of memory through gene therapy. That year one of the thrillers on the big screen was *Deep Blue Sea,* about super-smart sharks. Matt was tickled by the movie because his gene therapy project had strikingly improved the memory of laboratory rats. He liked to say that their IQ scores were 170 or so. *Very, very smart. A close-on genius rat.*

Matt was working on still other research projects so new and novel that he was having trouble getting the results published, he told Jamie. He thought Jamie might want to consider not only the EAAT2 gene therapy idea but also at least one or two of Matt's other, more unconventional ideas. There was one in particular about which Matt was very excited, a radically new approach to incurable brain diseases. He was developing a brain vaccine. It was extremely exciting—revolutionary, he said. He was preparing a paper about it. The vaccine was an idea that he had developed in New Zealand. In a way, it fit with the zeitgeist. There was so much interest everywhere now in the power of the brain and mind to control diseases. Most of that sort of therapy was soft science. But he had gotten interested two years before in teaching the immune system a few new tricks. He and his colleagues were looking for a way to train it to generate an immune response to a brain protein. The immune system can be very specific: That is, it can recognize a highly restricted range of targets, and attack only those targets. That is appealing, Matt explained, because in diseases of the brain it is difficult to find drugs that are so specific. Psychiatric drugs, for instance, are notorious for their side effects. So Matt and his collaborators had decided to teach the immune system to try to treat what the immune system normally ignores.

This idea is radical, and most of the early reviewers of Matt's paper thought it was impossible. For one thing, the brain is immunoprivileged: It is normally ignored by the immune system. And normally the body generates antibodies against foreign invaders, not against its own molecular machinery. But Matt's brain vaccines would have to generate antibodies against molecules that the brain itself had made, such as neuroreceptors or bad EAAT2 pumps.

However, Matt believed that he could do this. He could teach the immune system to interact with a problem protein in the brain and either clear it away or stop it from doing more harm. He had already tried this on rats, in an experimental vaccine against stroke damage. It is well known that when the brain suffers a small stroke, the body switches on genes that help protect it against the damage. The body is now preconditioned and can deal better with damage from the next stroke. Much of the damage is caused by a local flood of glutamate—the same problem that Rothstein argued was central in ALS. Matt and his team had devised a vaccine against a set of brain receptors that are sensitive to glutamate, the NMDA receptors. They had vaccinated the rats and then given them drugs to induce strokes. The vaccine seemed to have preconditioned the rats and helped protect them against the damage.

That year, the Human Genome Project was racing ahead and it gave hope to almost everyone in biology, from the most conservative to the most radical, that the new age of molecular medicine might be coming soon. When the project was done, humans would know the location and the code of every single human gene. The more we know about our genes and what they make, the more we understand where the making and remaking goes wrong. As the Human Genome Project matured, Matt predicted, more and more drug targets would appear. Then the immune system, which is so good at recognizing foreign invaders, could be trained to recognize hundreds or thousands of the malformed proteins the body itself has made. The Human Genome Project would teach us more and more about our own molecular engineering, and Matt and others would teach the immune system more and more tricks. Some mismade proteins it would clear away like a scalpel. Others it would neutralize. Still others it would modulate so that the proteins and their cellular neighbors could work correctly again.

Matt was considering heart surgery for a first trial of his neurovaccine. Like a small stroke, heart surgery can damage the brain: Whenever there is an interruption in circulation, the brain loses some neurons. With the right neurovaccine, he could immunize people prophylactically and protect them against the damage.

Teaching the immune system to heal what it could not heal without our help—Jamie found the idea visionary and thrilling. He knew just enough to talk gamely about it after his year at the Neurosciences Institute, since Edelman had won his Nobel Prize for his work on the immune system. Jamie urged Matt to consider a patient with ALS for his first test of the neurovaccine, since protection against glutamate damage there could mean everything.

The lunch went so well that Jamie and Matt had dinner, too. Soon they were spending weekends sailing together, both talking at high velocity. Matt During's apartment in Philadelphia is just a block or two from his laboratory, and Jamie began staying there often on his drives back and forth from Newton to Baltimore and Philadelphia while they planned the genetic engineering of Stephen. Paola Leone liked the project, too; from her first meeting with Jamie, she was deeply moved by the Heywoods' crisis. Paola was as emotional and empathetic as Matt was pragmatic. Her office was decorated with photographs she had taken of Canavan toddlers, and half a dozen stuffed animals and scrawled drawings, presents from doomed little children. She shared her three-story Philadelphia row house with a cat, a rabbit, and two German shepherds, Gnocci and Bambi.

Jamie told Stephen's doctor, Bob Brown, about his project with Matt During. In Baltimore, he told Jeff Rothstein, the discoverer of the EAAT2 problem. Rothstein happily agreed to run the tests on ALS mice to see if an EAAT2 therapy did extend their lives. In Philadelphia, Jamie, Matt, and Paola planned to design and build a DNA package containing the EAAT2 gene. Gene therapists call this "the construct."

Once they had built the construct, they would inject it into the mice. If the DNA injections helped the mice, they would test the injections on monkeys to see if they did any harm. If the injections passed those tests, then Matt and Paola and their team at Jefferson would inject the DNA into Stephen.

Jamie had now heard through the grapevine that Margaret Sutherland, the young researcher at Vanderbilt in Tennessee, had crossed some of her mice, which made too much EAAT2, with some ALS mice. He was very curious to know what had happened there. But injecting the DNA into the spines of the ALS mice would be the most direct test of his idea that they could try.

Jamie, Matt, Paola, and some of the young scientists in their laboratory celebrated the plan at Paola's row house in Philadelphia, and they all got drunk. Jamie and Matt played game after game of pool down in Paola's basement. Jamie began trouncing Matt, and he could see that Matt was not taking it well. They are both competitive at everything they do, and Matt did not want to lose. Jamie liked seeing that. He needed a partner in this project who did not want to lose.

Matt won the next round of pool.

Jamie flew Matt up to Newton to meet his family. John and Peggy held a big dinner party, and all of the Heywoods feted Matt During.

{ TWENTY }

Moral Capital

Now that Jamie had put his dream team together, he had to raise money. The whole family pitched in again to plan what Jamie billed as the First Annual Belly Dance Extravaganza. (That "First" was typical Jamie.)

I once watched Melinda belly dance at Karoun, a Middle Eastern restaurant in Newtonville, the place where she earned her way as an undergraduate at Wellesley. Melinda danced out, clinking finger cymbals, wearing a costume of strings of coins and a pleased smile. The smile was not exactly a come-on smile. It was the smile of the dance.

I was sitting at a table with John, Peggy, and Jamie. "Do you know which one she's doing tonight, Peggy?" asked John, in his proper British accent, looking absolutely comfortable and at home.

"Candles, I think."

Melinda laid a tray of burning candles on the floor. Then with a flourish she set it on her head and began to gyrate and undulate to the music. She looked very beautiful.

"Well," said John, "she started out as a young child, dancing with her mother. And danced pretty steadily since. Well, you know, trumpeters earn their way through college playing trumpet, and belly dancers earn their way through college belly dancing."

Melinda left the dance floor and came around the tables for tips.

She gave Jamie a few knocks with her hip as she went by, explaining to the next table, "I know him."

"Melinda's very good at this," said John. "She keeps it very light-hearted and funny."

Jamie seeded her sequined girdle with a twenty. John inserted a ten. "*We* don't get it back," Peggy explained.

I inserted three ones.

"Excellent technique," said Melinda.

She and her sister Piper had been dancing for drachmas and dollars since the age of seven. They said they would be happy to shimmy for the Heywood family cause with their trays of burning candles, their jugglers' torches and impossibly balanced swords.

"Beauty, brains, and a back," said Melinda.

Down in the basement, where they planned the extravaganza, Jamie paced between the door-desk and the workbench and delivered speeches about the mission of the new foundation, and what they were about to make possible. Usually his audience consisted of Melinda, Peggy, and Robert. Stephen and his carpenter's assistant kept passing through, tromping up and down the stairs.

Stephen was still working on his parents' master bathroom. He had expanded the project to include a home office for Peggy at the landing at the top of the stairs, with skylights at clever angles to let in maximum light. Like Jamie, he always had a way of drawing high-powered people into his projects. His carpenter's helper this time was a two-star marine general with Purple Hearts from Vietnam, a retired man of about sixty with a bit of a limp. Since retiring from the army, John Brickley had served as director of the American Lung Association in Massachusetts. Stephen hired him in the middle of all this action on their home front. Brickley said he wanted to learn carpentry from a pro. Stephen paid him the standard wages of a carpenter's apprentice. He and the general were a match in manly taciturnity and neither ever talked about the other's limp. Whenever any of the Heywoods told Brickley how much they appreciated his help, he would shrug. *There's only so much golf you can play.*

Jamie's speeches in the basement were also typical Jamie—raising the ante. He barely had his ducks in a row, and he and Melinda were agonized that they were paying Robert's salary, and buying a two-thousand-dollar printer, and living in the house on Mill Street, all on his parents' money—while slowly running out of money themselves. But Jamie foresaw an organization that would work like a machine to cure one incurable disease after another. They could do it. He could see the whole thing in his head. Again and again he explained that ALS is an orphan disease, too small and too baffling to be worth even a minor investment by a pharmaceutical house. Merck or Pfizer would not gamble the time or money on basic research in ALS: too long a shot for too small a pot. But if, with a small investment, Jamie and his team could show that his EAAT2 gene therapy idea or Matt During's neurovaccine was promising, then a biotech like Genzyme *would* be interested. Then they might spend a few million dollars to develop it.

"Contrast ALS with heart disease," Jamie would say. "This stroke stuff of Matt's—when that paper of his comes out, people are going to be clamoring at the door to get it. The neurovaccine is going to be worth billions, so an investment now of a few million is totally worth it."

Jamie was dreaming there, pumping himself up and pumping his troops. He did that often. Of course, it was impossible for any of them to know what might be about to happen and what would never happen. So much more had already happened than any of them had expected when Jamie came home from California. When you listened to Jamie, it was easy to forget that gene therapy had not yet saved a soul.

He was thrilled by the idea of combining a nonprofit foundation with a corporate spirit. As a desperate entrepreneur, not a scientist, he was sure that he could make science race ahead. "Research is slow," Jamie would say. "Development is fast." They would yank EAAT2 out of the doldrums of academia. "You get it out of the hands of PhDs and into the hands of technicians, who actually *work for a living.*"

That was pure Jamie. It was also typical of the tension between activists and scientists. Activists want action. Scientists want knowledge—for its own sake, and for the infinite and unpredictable ways that knowledge can help lead to action. Not long ago a famous biologist told the actor and activist Christopher Reeve, "You can never do enough basic science." The Christopher Reeve Paralysis Foundation no longer supports that biologist.

"I'm just saying all these things for the first time out loud," Jamie would say, for the tenth time, pacing up and down in the basement, and building castles in the air with the money they hoped they were about to raise. He foresaw a staff of roving scientists who would read everything new in the literature, like Joe Gally, and go out to talk to people in every field on the borders of ALS, hunting for clues. "These are the kind of things that science is both good at and bad at. Epilepsy researchers don't talk to ALS researchers. It just doesn't happen. They don't go to the same conferences."

He harped on epilepsy because he was thinking of the young researcher Margaret Sutherland in Tennessee. By now the litter of mice she had gotten by breeding ALS mice and the mouse she had engineered, a mouse that made extra EAAT2, must be growing up. But she was not talking about them. She was a young scientist, and it was the kind of side project that could go on for years, without a publication in the scientific literature and without ALS people ever even hearing about it.

Jamie saw himself as the impresario of a band of scientific operators. He would be a sort of Robin Hood of WIIF'M, convincing the rich to give to the poor, winning by brains, charm, cunning, moral capital, and super salesmanship. "There is all this amazing stuff going on out there in science," Jamie said. "We're acting here as a moral venture firm. We are a venture firm operating in the interest of our shareholders—which is the survival of the patients who have ALS now. Moral capitalism! That is one of our greatest resources, our degree of moral capital! How much do we have in the bank? We have a lot in the bank! Basically we have unlimited funds in terms of moral capital."

The spring of 1999 was the perfect time for a millennial venture. Technology start-ups were raising ridiculous amounts of money. The tide that had raised real estate values in Palo Alto and made Stephen's cottage sell for almost a million dollars was now lifting all boats. On the stock market, eBay was trading for almost 4,000 times earnings. The ratios at Amazon.com and Priceline.com were even better. Their ratios were infinitely better, as one market cynic pointed out a little later, since their stocks were skyrocketing and they had no earnings at all. In Silicon Valley, the founder of Netscape was now dreaming of swallowing up the whole health-care industry in a sort of giant Pharma.com he called Healtheon.

Investors with stakes in the Valley were looking at biotech. Wall Street put six billion dollars into biotech that year. That was six times more than the Street had invested the year before. It was sixty times more than investors had thought the industry was worth ten years before. The investment frenzy fed on itself. Leaders of biotechs could turn around and use investors' mania to raise more money and draw new recruits. William Haseltine, the director of Human Genome Sciences, an apostle of regenerative medicine, gave a talk to grad students at Johns Hopkins that spring titled "A Coming of Age." "Accompanying the exploding interest in genomics," he said, "our stock price has nearly tripled in the past six and a half months."

Melinda still danced at Karoun one night a week, and the owners treated the Heywoods like family. They turned their place over for the First Annual Belly Dance Extravaganza at cost.

"My parents were a little nervous about asking for money from their friends and stuff like that, but they did," Stephen told me a little later. "I know that's been quite hard. Also, my dad asking his colleagues I think was quite difficult." But John and Peggy each wrote letters to friends and family, and Jamie and Melinda and Robert called everyone they could think of. At the same time they encouraged Stephen and Wendy to live their lives. They did not want them to lose their time alone together. The two of them were now deeply absorbed in each other and trying to forget how short a time they might have. On the night of June 1, 1999, Wendy wrote in her journal:

> One-half hour of the first day of June left. Nine months with Stephen, nine wonderful months. I knew I had love in my future, but I never dreamed I could be this happy.
>
> I stop sometimes, my foot drags just so in the middle of a quick step, my stomach drops and silently I plead, *No.* I've made so many deals with God of late I am up to my eyeballs in religious debt. It is impossible for me to believe I would be without this most wonderful, giving, sweet man. How lucky I am to have this incredible man in my life.
>
> Sometimes I want to rush things, marry him, have children. Do it now. But I want him to be happy with his timeframe for his plans. I know he has our future mapped out in his head and it's a route I'd never doubt. I trust him. And, the rushing feeling always follows some discussion about ALS. Although part of me wishes I had a ring to wear for my visit home. Christ, I feel like such a girl for writing that!

On the night of June 3, 1999, every seat in Karoun was taken by scientists, Heywoods, honorary Heywoods, neighbors, well-wishers, congregants from Grace Church. Jamie made a speech praising the extraordinary scientists in the room. He explained that he was concentrating on gene therapy because that was the fastest way he could get a treatment into Stephen. He spoke with honorable realism of his chances—more realism than he normally allowed himself. But he was candid about his need for optimism in the face of his chances. "I've struggled with this. A few months ago—well, I say that, but actually it was only a few weeks ago—I made the decision to believe that we can do this. We can beat this disease.

"Time is the only thing that matters," Jamie said. "I cannot make this clear enough. Time is the only thing I cannot go out and get more of.

"Now I'd like to hand this over to my brother, who promises to lighten it up."

Stephen stepped up to the microphone. He was two weeks away from his thirtieth birthday. He still looked very fit, taller and sturdier than Jamie. He told the crowd that he thought Jamie's enterprise was like an Internet stock. He meant that it was young, innovative, glamorous, and they would all enjoy spectacular returns on their investments.

He and Ben had always been movie freaks, Stephen said. "We don't necessarily like *good* movies. We like *Hollywood* movies. And one thing Hollywood is great at is a tearjerker. The scene that always gets me in a Hollywood movie is when someone dies to save someone else's life." Lately, Stephen went on, he found himself imagining a situation like that more and more often. In his daydreams he died a heroic death, rescuing a family of five from a fire. "Except I'll be in a wheelchair, so it will have to be a very slow fire, in a house with ramps." Sometimes he wondered why he was daydreaming like that. When he tried to explain it to himself, he said, other clichés came into his mind, like the old line, "You have to live your life like every day is your last. Which I think is ridiculous," Stephen added, "because you'd be hungover every morning, and no one would ever take out the garbage." Still, Stephen guessed that he was giving himself that Hollywood ending because he wanted his life to have meaning. "Another cliché."

Jamie had given him a great gift, Stephen said. "It's the ability to live my life, with my friends and family, as I see fit, while knowing that my life will have meaning. I know that my illness will save other people's lives. So, Jamie, thank you.

"You all know my family," Stephen said. "Many of you know me. I personally guarantee that out of this will come a breakthrough, a treatment, or a cure."

Stephen made his way back to his table, and Robert Bonazoli took the microphone. "There's an old saying, 'If I can't dance, I don't want to be a part of your revolution,'" Robert said. "Well, if there's one thing we have in our revolution, it's people who can dance."

Then the Daughters of Rhea, Melinda and her sister Piper, danced out with finger cymbals. Melinda did her candles routine, and Piper whirled and reeled with a scimitar balanced on her head. Robert announced, "Tonight only, all belly dance tips are tax deductible." And crowds of Heywoods and their friends stuffed checks and bills into the sisters' sequined girdles. Then Melinda and Piper led the whole crowd in a chain and they snaked around and around through the brick arches of the restaurant.

Stephen leaned back in his chair against the white brick wall, holding Wendy's hand, and watched the chain of dancers. He looked almost foreign that night at the edge of the festivities, as if he had traveled such a long way to be there and brought such a different world of emotions and associations from his own country that now just being there was all that he could manage. His expression was somber, pale, serious, and spent.

{ TWENTY-ONE }

Birds Suddenly Appear

With the tips the Daughters of Rhea collected in their girdles, and the checks that came in the mail the next week, including a generous donation from Alexander d'Arbeloff, the First Annual Belly Dance Extravaganza raised $240,000. That was a pittance in a normal research setting, but it was close to the amount the family of the Canavan baby Jacob Sontag had donated to Matt During's Canavan research, and Jamie hoped it would be just enough for him to begin his mad dash in the style he had in mind. A quarter of a million would go further for him than it would for any medical research charity he could think of, because Jamie would not give the money to anyone in a lump sum. He would parcel it out. Now that he had put together a dream team, and a little money, he planned to run the show.

In Matt's office he sat in on the lab meetings and more and more often he ran them with Matt. Like gene therapists all over the world, Matt and his team were struggling with tough engineering questions. A gene has to work as part of a system, like a gear in an engine or a sentence in a paragraph. To make it work, a gene therapist has to engineer what is called a promoter sequence, which sets in motion the reading of the gene. Otherwise the gene will get into the nucleus and sit there unread. The gene therapist also has to provide a stop codon, like a period, to get the cell to stop reading the gene when it comes to the end of the message. And a gene therapist also tries to include a switch that can be turned on and off from outside. That way, if the gene they have injected is hurting the patient, they can turn it off before it does too much harm. In experimental gene therapy on flies, temperature has been used as a switch. The new gene turns on only when it is very warm. At room temperature the gene turns off again. In human beings, certain antibiotics can serve as the switch.

Taken together, all this DNA—a long strand containing the promoter sequence, the gene itself, the stop codon, and the switch—is called the DNA cocktail. It is the elixir of the gene therapist. To get it into a patient's cells, the gene therapist engineers a virus that will carry the elixir instead of its usual load of DNA. A single virus is like a syringe. It has evolved to be very efficient at injecting its DNA into a host cell so that the DNA will snake through the cell and find its way into the nucleus and insert itself into its host's DNA. With the standard tools of genetic engineering, it is now a simple job to take a sample of viral DNA and cut away much of it, and put back the harmless bits that give the DNA the ability to snake into the heart of the cell. It is also simple to splice in a new ribbon of DNA—the elixir, the DNA cocktail. But the head of a virus, even a hollowed-out virus, is a very small syringe. Gene therapists usually find that they have more DNA in their cocktail than they can stuff into the virus.

Essentially, a virus is a Trojan horse. Its contents are DNA that tricks its way inside the city of the cell, and once there, gets in through a nuclear pore to the DNA in the center and takes over. Then the cell makes more of that viral DNA. Once a cell has been infected, it literally explodes and hundreds of syringes spill out to infect cells around it. What a gene therapist does is to scoop out some of the Greeks from the Trojan horse—remove all of the most dangerous ones—so that the horse really becomes a gift to the city. The trouble is that viral DNA, after innumerable generations, has evolved an exquisitely compact force inside the horse. That is why there is so little room in there, and that is why the potions of gene therapy tend to be too big to fit inside that tiny hollow space.

That same spring, for instance, while Jamie began meeting with Matt to plan the EAAT2 project, a parallel project was in the works at the University of Pittsburgh. There a gene therapist named Xiao Xiao was working on gene therapy for a form of muscular dystrophy—the most common form, Duchenne muscular dystrophy, which afflicts about one in every 3,500 male children. Their muscles waste, they are confined to wheelchairs by the time they are thirteen or fourteen, and they are dead by their early twenties. Doctors had no treatment to offer, and no cure. Biologists knew the complete sequence of the gene that causes the problem—the dystrophin gene. Inside a human cell, in the form of messenger RNA, that gene is about 14,000 letters of genetic code. Gene therapists had been looking at that gene for a long time, but they could not fit so much DNA inside the head of a virus. There was room in the viral capsule for only about 4,000 letters of code.

In Pittsburgh Xiao Xiao and his colleagues studied the shape of the molecule that is manufactured by a muscle cell with a healthy dystrophin gene. It is a long semirigid molecule: a rod with sockets at either end. The sockets are important. They are known in the jargon as functional domains. One socket binds to the plasma membrane of the muscle cell, and the other to the cell's cytoskeleton. Without them the membrane is not properly attached to the cell's skeletal scaffolding. A muscle cell gets a lot of wear and tear whenever the muscle flexes. Without those rods, the membrane comes unstuck and the cell falls apart.

The rod consists of twenty-four protein subunits like twenty-four LEGO pieces, all about the same size and shape. It also has four hinges, spaced along its length, which give it flexibility. The rod has to be flexible because it has to give gracefully every time the muscle flexes; otherwise it will be torn apart and the cell will die.

The gene for this simple piece of molecular hardware is on the X chromosome. Girls have two X chromosomes. Since they can make these rods even if they have only one good copy of the dystrophin gene, they do not get Duchenne muscular dystrophy. But boys have one X chromosome and one Y chromosome. The Y chromosome is small and atrophied, and it has lost its copy of this gene—and many other genes, too. That is why boys inherit more genetic diseases than girls. If a boy inherits a defective copy of the dystrophin gene on his X chromosome, he will make none of these rods, and his muscle cells will slowly fall apart.

This is a natural project for a genetic engineer. Why should a boy die for lack of such a simple piece of hardware? The rod is not much more complicated than those braces they sell at Home Depot to screw into freestanding bookcases and keep them from falling down. If the boy's DNA has forgotten how to make those rods, why not teach it how?

Xiao Xiao considered the rod: a long and and more or less featureless repeated assembly of sections, with the four hinges for flexibility and the sockets at the ends. Then Xiao engineered a radically abridged version of the gene, cutting it down from 14,000 letters of genetic code to 4,000. The abridged version of the gene made a shorter rod. It had the same sockets at either end, so he thought it could still attach the membrane to the cytoskeleton, as if holding Sheetrock to the frame of a house. And it was still flexible, because Xiao Xiao had kept two of the four hinges, so it might still have enough play in it to last. In diagrams, the object that a cell made with this reengineered dystrophin gene looks like a prosthesis, like an artificial arm or leg, not as flexible as the original but better than nothing, a thing worth having if you need it.

Xiao Xiao knew that nature had already tried this engineering experiment. There are boys born with mutations that give them short rods. The rods hold that membrane to the muscle cell cytoskeleton, and those boys develop only mild forms of muscular dystrophy. Other researchers in the field of Duchenne muscular dystrophy had found short rods in the muscle cells of a man who had lived for decades with a mild form of muscular dystrophy. He was in his late sixties, and he could still walk.

So Xiao Xiao inserted his abridged gene into a viral capsule. He and his colleagues cultured their engineered virus in vats. From Jackson Laboratories in Maine they ordered mice that had been genetically engineered to get muscular dystrophy. The mouse dystrophin gene is defective and makes no rods. At around three weeks old these mice begin to suffer, their muscle cells degenerating in great sick waves of trouble.

When Xiao Xiao and his colleagues were ready, they filled a syringe (a real, life-size syringe) with a milky solution containing more than 5 trillion virus particles per milliliter. They injected these 5 trillion virus particles into the hind-leg gastrocnemius muscles of forty mice with the defective gene.

And three months later, the muscles in those mice had not degenerated. Nor had their immune systems attacked the virus, or the rods, or the muscle cells. They sacrificed the mice and put tissue samples under the microscope. The muscles looked normal. Apparently the rods had attached properly at both ends and anchored the plasma membrane to the cytoskeleton.

Time would tell if the short rods would hold. Like artificial hip joints, they could pop out when mice or men tried to do something eccentric and frenetic. But this was the kind of hopeful new experiment that inspired Jamie in the spring of '99. There were many genetic engineering successes to study that year—in mice. Many treatments and cures that heal mice fail to heal human beings, so researchers with a success in a mouse lab can never be sure how hopeful to feel. Still, there really were some reasons for hope. A team at Boston's Children's Hospital and Harvard Medical School tried using stem cells to cure mice with muscular dystrophy. They injected bone marrow cells with the healthy dystrophin gene into a tail vein of the mouse, about as many cells as Xiao's team had injected viral particles into the gastrocnemius muscle. That experiment seemed to help. This was the very approach that Martin Cline had tried and failed with back in 1980, in the world's first attempt at gene therapy.

Jamie's plan was real now, and Peggy felt the need to honor it. From her years practicing psychotherapy in Newton she knew how many families fall apart around troubles like these, and she was determined to hold hers together. And the foundation was beginning to mean as much to her as it did to Jamie. Jamie was always putting her to work. For a time he had Peggy cold-calling people on their mailing list for donations. But she was getting turned down, and she came to him weeping. *Every* no *is like a nail in Stephen's coffin!*

Jamie switched her to something else: writing thank-you notes to donors and well-wishers.

That month, a white, peeling, slightly dilapidated Victorian went on the market in Newton Highlands, just a few minutes' drive from the Heywoods' house on Mill Street. Stephen had long since given up his dream of renovating a house in Boston. But the house and its garage, a ruined carriage house, was just the kind of Boston fixer-upper he had dreamed of. "I looked at this place and, suddenly, *bang!*" Stephen told me. "I'd assumed I wouldn't work anymore. But now this—it was perfect."

Peggy knew that buying the house would mean mortgaging the house on Mill Street and cutting into their retirement savings. And then there were the expenses of Stephen's illness itself to consider: The expenses were bound to grow exponentially in the next few years. But Peggy wanted to give that house to her sons. She knew that Jamie and Melinda needed a place to live outside the house where Jamie had grown up. And the foundation that Jamie was putting together needed a way to get out of the basement. And the carriage house might be right for Stephen and Wendy someday, though it was hard to plan ahead that far since they did not know how long Stephen had to live and since Stephen and Wendy were not even engaged.

The house—even dilapidated as it was—would not be on the market long, and John Heywood was out of town on one of his consulting trips.

Peggy thought, and, with contributions from Jamie and Stephen, she bought it.

"Thank God I'm not married to a flaky man," she says. "Thank God John's John."

It was characteristic of John Heywood to take this in stride. Sometime after he got home, he mentioned his book *Internal Combustion Engine Fundamentals,* and said he hoped to bring out a new revised edition someday. "A retirement project."

"If you ever retire," Peggy said. "You may never retire now."

So Jamie and Robert moved the door-desk out of the basement and installed the new foundation on the ground floor of the Victorian. They set up their desks in the living room, and their computers, and homemade inspirational signs and slogans. Jamie tacked a sign up on his bulletin board: "Depression is merely anger without enthusiasm." On a bookshelf he propped a saying of Melinda's in a little gold frame: "I'm halfway between euphoria and falling on my head. Melinda Heywood 6/2/99." Jamie and Melinda moved into the second floor of the house. Someday they hoped Stephen would fix up the carriage house. Jamie would save and restore Stephen from the main house, and eventually Stephen would restore the Victorian from the carriage house.

In July, at the Heywood family retreat in North Carolina, Stephen's right arm and right leg were so weak that he was forced to sit out the annual basketball game. Jamie tried to hide his feelings from Stephen. This was the loss he had dreaded in the first hour of Stephen's news. By now Jamie could see that giving up basketball would be the least of it. Stephen was even having trouble eating.

"We have a big crab fest at the beach," Jamie says. "Bushels and bushels of crabs die for our family. And it was the height of brotherly love to actually crack open crabs for your brother and not yourself."

It was a big summer at Duck. Stephen smuggled a ring down for

Wendy. He had sensed a dozen times that year that she was expecting him to propose, but he had not. "I think I may have been waiting to figure out if basically she didn't have any choice in the matter," he told me. By now he felt that their romance was beyond his control or hers. The choice was out of their hands. Their love life was rushing even faster than Jamie's project. They were truly, madly, conventionally in love, only Stephen was dying.

"It didn't matter what I thought was best for her," Stephen said. "We were gonna do it anyway. And then it was just a question of shopping for the ring."

He had some fun with that. Early that spring, Wendy had gone shopping with Melinda and found a ring she liked. She had told Stephen all about it. The price was $1,800. *I hope that's what you had in mind,* she said.

Well, actually, more or less double that.

Oooh, said Wendy.

Then the two of them had gone shopping for a ring together; and when Stephen was sure that he knew what she liked, he picked one out in secret. *It's an antique, platinum, it's gorgeous,* he told his friends. But he told Wendy, *I'm still looking for a ring.*

Each summer at Duck, the Heywoods round up all the little kids in the family and hold a talent show on the beach. Stephen decided that he would get his brothers and cousins to sing a corny old song to her, "Why do birds suddenly appear . . ." While they sang, he would draw her up there with them, and in front of the whole family he would propose. Stephen had it all planned out, but an hour before showtime he realized that it might not be the best idea to do it like that. He decided that it was not a good idea at all to spring it on her in public. Next he thought they would go walk on the beach, and he would ask her down by the ocean. But it started raining, and they were trapped with a mob of friends and relations in the beach house that John and Peggy had rented that summer, a great three-story place with the seaward end on stilts. The talent show was only an hour away. *How am I gonna do this? We can't go for a walk on the beach, it's pour-*

ing rain. Finally Stephen just led Wendy out the door and proposed right there, pulling her under the house to get out of the rain.

They went inside and told his family. And later, at the talent show, Stephen surprised Wendy when he dropped to one knee in front of his brothers, his cousins, and all that summer's honorary Heywoods. He proposed to Wendy again, while Jamie, Ben, and a few of the others serenaded her, hamming it up and wavering way out of key.

Why do birds
Suddenly appear
Every time
You are near?
Just like me
They long to be
Close to you!

{ TWENTY-TWO }

Fizzy Water

In one of Jamie's late-night phone calls that August, I asked him if he felt more optimistic or pessimistic now that he had gotten the doctors to listen to him and work with him.

"I alternate. I'll give you a quick antidote," Jamie said, without noticing his slip of the tongue. That past weekend he and Melinda had gone canoeing with Stephen in Newtonville on Crystal Lake. The three of them were carrying the canoe to the water, Stephen and Melinda in front and Jamie in the rear. Jamie was feeling terrific at that moment, he told me. He had yet to read of a single drug that had not taken six to eight years to get to clinical trials. Now he had a chance of getting his gene therapy into Stephen in a matter of months. He thought, *My God, it's unbelievable what we've gotten done!*

Then Stephen stumbled, dropped his end of the canoe, struggled ahead along the path in slow motion, and collapsed.

"I—" said Jamie over the phone. We both waited while he recovered his composure.

"This thing is relentless," Jamie said. "You panic about hours. Sometimes you think: 'This is unreal, the progress we're making.' Then you think, 'It's not real enough.' "

Jamie dropped by my town, and we met at Paganini once again. What a difference. He and Robert Bonazoli were on their way home from another long day's journey to Jeff Rothstein's lab in Baltimore and Matt During's lab in Philadelphia. In Philadelphia they had treated Matt During to a long, late lunch at the Striped Bass. From the way that Jamie said the name, I could tell that it was a hot place.

Jamie and Robert both wore neatly pressed long-sleeved shirts and khakis, and they looked very young to me. Robert's brown hair was longer than Jamie's and windswept. He wore stylish wire-frame glasses, a small silver earring, and sandals. The waitresses in the café kept looking at both of them.

"We held a foundation meeting in Dad's lab a few days ago," Jamie told me. It was a scientific seminar about ALS. He called it ALS 101. Not the usual project that got talked about at the Sloan Automotive Laboratory, Jamie said with a laugh. "It's a '50s, engineering, mechanical kind of building. And we took it over for two days." The leader of the seminar was Jamie's mentor and old office mate, the Walking Library of the Neurosciences Institute. For a few days Joe Gally was living his dream: He was teaching molecular biology at MIT, and his students were listening as if what he had to say meant life or death.

Jamie and Robert told me they were trying to rate all of their options in terms of risk and reward. They talked about Jamie's gene therapy project and Matt During's idea for a neurovaccine, his trick of getting the immune system to heal the brain. During's idea was so novel that he had still not gotten it published. Reviewers of the paper kept insisting that he do more experiments to prove that his technique really did what he claimed it could do. Jamie explained that unlike gene therapy, where each treatment has to be meticulously fitted to each disease, a neurovaccine could be adapted quickly to different neurodegenerative diseases. He had persuaded him to try out a vaccine to treat ALS. "The way I think about it, it's a whole new kind of medicine," he said. "Like gene therapy or stem cells. A whole new way of manipulating the body. Very exciting and potentially very powerful. This is why Matt is such a genius."

The gene therapy was moving along fast now, Jamie told me. "The experiment started today," he said, glancing at his watch. Jeff Rothstein, the ALS expert at Johns Hopkins, had started injecting ALS mice with some genetically engineered viruses from During's lab. "This is exciting," Jamie said. "The work would have gotten to ALS *eventually,* but it wouldn't have happened for a long time."

If Matt During could make a vaccine against ALS, I wondered if he could also make a vaccine against Lewy body dementia. But I kept that thought to myself. In those days I did not talk much about my mother, even with Jamie.

"I don't know if the neurovaccine will work," Jamie went on. "It's very clear that the gene therapy project is still the best we've got going. The vaccine's ambiguous. I'm excited, but I'm worried it might be toxic. I've also got seven to ten other ideas."

And he did—but as Robert confessed to me later, those were all so much window dressing.

I told Jamie and Robert that I was amazed how much they had accomplished.

"At this point, we're three folks in a living room," Robert said wryly.

"Better than a basement, though," said Jamie.

I looked at them, out on the sunny deck of Paganini, drinking bottled water. The image of the Internet start-up seemed about right. They were young, they were excited, they were frankly selling their youth and excitement, and it really was hard to judge how much else they had. They were a brand-new start-up, and they entertained me with the kind of hyper sales patter that was circulating in 1999 in Cambridge, Boston, San Francisco, and Seattle, the rhythm of the charming sell. Every few minutes, Jamie checked his cell phone.

"Why don't I hear it ring?" I asked, feeling like a rube. Jamie was an early adopter and I was a late adopter.

"It's set on vibrate," Robert explained. We both watched Jamie listen to his messages. "When Jamie doesn't know what else to say, he says, 'The only thing we don't have is time. The only thing we don't have is time.' And it's true," said Robert.

Jamie slipped his phone back in his pants pocket. "Another slogan," he said: " 'Our responsibility is to spend money irresponsibly.' "

"This is war," said Robert. "The enemy is disease. To win you need all different kinds of missions and platoons. We're like psycho commandos doing really wild stuff. We could fail. But if we hit, we're going to hit big. The doctors are more conservative."

"And they should be," said Jamie. "They're playing for the long haul."

"We're trying to make it happen fast."

"Make it happen fast for Stephen."

"We're a venture capital firm," said Robert. "The model is exactly the same. But instead of raising money to give away to scientists, we're raising money to try to cure ALS ourselves. It's so simple."

I doubted that "it's so simple" but let it go. Jamie was giving me his sell-you-the-moon stare. He had a salesman's way of holding eye contact with a level gaze and beaming pure energy. "I come from an environment where you figure out how to make someone do what you want them to do," he said. "It's not that I'm a biologist, or a doctor, or a rich man. But I'm enough of all those elements now that we can make something happen. Never underestimate the power of hubris. I'm too stupid to know not to try. There's a certain power to that. Just get in there and start swingin'. No reason why we can't have therapy fast. Just got to move mountains. So what? Mountains can be moved. They've been moved before. Hopefully I can blow you away *again* in two more months."

"Jamie says, Keep setting impossible goals, and keep meeting them. We *said,* Let's raise two hundred thousand, and we *got* two hundred thousand," said Robert.

"Helps to have moderately wealthy friends—and a lot of them," said Jamie.

"We made a plan and we did it," said Robert. "And by the way, I don't have any moderately wealthy friends. OK? Don't include me in that."

Jamie ordered us three more bottles of fizzy water.

"Working up quite a bill in fizzy water lately," said Robert.

If emotional intelligence is the ability to feel many sides of a question at the same time, as Daniel Goleman says, then by the time the shadows lengthened in the street outside the café, I must have had a very low IQ. I forgot my reporter's objectivity. Jamie and Robert had charmed me so completely that I forgot myself, and the sick, and the life-and-death questions that lay ahead for all of us. All I knew was that they were offering me a wonderful story on a charmed summer day, and I was ready to join their adventure.

It was easy to feel that way—and it was just as hard to force oneself to stay realistic. For Jamie and Stephen that had to be hardest of all. Jamie told me that Stephen and Wendy had decided to put the Cave behind them and get a place together. They had found a beautiful apartment on Beacon Street, in Boston's Back Bay. Jamie could not understand it. The place they had picked was a fourth-floor walk-up.

By the end of dinner at the café, Jamie seemed to think he had sold me on his story almost too well. "Still," he reminded me at last, "we're painting a pretty picture here. There's really hard stuff going on. My brother is probably going to die."

I wanted to deny it.

"That's reality," Jamie said. "He's someone I love dearly, someone I am very close to. That makes these things hard. Rob can tell you. It's going to be murder to go through as it happens. So I don't want to paint too pretty a picture. It's very scary and hard. I don't mean in a business sense. When I saw Stephen collapse! You just want to go put your fist through walls."

When I got to know Jamie better, I realized that he very rarely mentioned the possibility of Stephen's death. In fact, I do not think I ever heard him mention it again. Once, much later, I asked him if he had really believed what he said to me that day at the café. "What your rational mind knows and what your heart knows are different things," Jamie told me. "I think I knew rationally exactly what the chances were, and every time I sat down and thought about it I went to that correct understanding. And I don't think I ever deluded myself. The other side of that is, I haven't had a problem I haven't been able to fix yet. I've managed to fix everything I've tried to fix in life, whether it be a house or a machine or a design. So I knew the chances were close to zero but I also haven't missed yet.

"I think that's a constant internal struggle. It takes hope to dream about opportunity. And yet you need to keep yourself in the real world all the time. I knew the odds were incredibly slim. Yet I was still hopeful. It's that combination of realism and hope that makes you go like crazy for those small odds."

"Look at the scientific process," Jamie said at the café as we finished the last of our coffees. Somewhere in the course of his race he had made time to read my books and he knew how different this story was for me. The scientists I write about spend decades, lifetimes, on their projects. "And we're saying we're going to do it in months," Jamie said, and shook his head. Pensively he stirred the ice in his cappuccino with a straw. "It's invigorating and it's kind of terrifying. Sometimes I get to Friday and I just melt. Just gone. What I do is try to keep fifty balls in the air. And one hundred scientists. It's a mess, a total mess."

One hundred scientists? I should have doubted Jamie again. But he knew how to sell even when he was fretting and being realistic, which is a useful skill for an entrepreneur and impresario like Jamie. Looking back, I wonder if as a superb salesman of the near-impossible, he was anticipating my skepticism. He thought he was talking to an entirely rational mind. He did not know he had sold me so well that I was not feeling skeptical at all.

"Some might say our project is hopeless," Robert said, in the same tone as Jamie. "A skeptic would say that Jamie is tilting at windmills like Don Quixote. But the difference is, this *could work.* So you ignore the odds."

I felt elated: drunk on hope, admiration, and fizzy water. Jamie and Robert had charmed our waitress, too. She was about twenty. She came back after we had paid the bill and stood at our table looking at Robert.

"Well," she said. "Wherever you're going tonight, whatever road you choose, and however that road may twist and turn along the way, I hope you have a wonderful ride."

"Thank you," Robert said.

But she did not leave. She kept on looking at Robert.

"Where *are* you going?" she asked.

{ TWENTY-THREE }

Lead into Gold

The next morning, Stephen and Wendy flew from Boston to Paris. Ben and Melinda went with them. Jamie took care of a little more business; then he joined them, and they all had a good time. There was one bad moment when Stephen stumbled on a long flight of stone steps at the Louvre. It was like my mother's fall on the steps of the Met. But her falls were sudden: the marionette with the snapped strings. Stephen had a way of falling in slow motion. Even when he was walking on a level he would keel very slowly off balance while struggling hard to right himself over a space of twenty-five feet. Sometimes Jamie thought Stephen was almost funny to watch—when he was not on a flight of stairs.

There was another bad moment in Paris when Stephen challenged Melinda to arm wrestle. She knew that he wanted her to try her hardest. Feeling crappy about it, she forced his right hand down to the table.

While they played in Paris, I took a train to Philadelphia to see the gene therapist Matt During at Jefferson Medical College. On the way down I thought about the chance that I might do something for my mother. I love my father dearly, and we are very close. But in the first few years of her illness he could be exasperating. First there was the battle to get him to take my mother to a psychologist when her personality darkened and she began to brood. To him that sounded like a shameful step. When I pushed him, he always gathered himself up into formality, and I went from Jon to Jonathan. "Well, Jonathan, it's up and down, your mother goes up and down. This week she seems a little better." How many times we had that conversation! "Well," and that repressive pause, "Jonathan." As if to push me back into the rows of chairs in a classroom—to make some room between the lecturer and his audience of one. He said the thing to do was wait.

And when she began her rag-doll falls, how long it took for him to agree to put in a railing on the front stairs and remove the throw rugs she kept tripping over. At least that was how it seemed to me. Everything was so slow, so slow. He was holding on—each step would acknowledge her decline and the change that had come to them, and he was holding on to the way things were. Just to roll up a rug would be to begin dismantling the beautiful house she had created when they first moved in. He did not want my advice, and he did not want anyone else to know what was happening to her. In summer even on the hottest days I could not persuade him to open the curtains, the shades, and the windows.

(My father saw it all differently, of course—including the windows. That was an engineering decision, he told me. His best friends in the department did the same thing. I did not understand the concepts of heat transfer and thermodynamics.)

But now she was sinking fast and he was at least as desperate as I was to find help for her. It made a difference for my father, I thought, that we had a diagnosis now—Lewy body dementia, something concrete and physical, something in the realm of engineering and mechanics, not the shaming realm of mental illness. And if Matt During's neurovaccine looked like our best hope, I thought he might agree to bring Ponnie in to a laboratory and have her injected with an experimental treatment from the edge of medicine.

I had stopped writing about my mother in my journal. Later I found out that Melinda had quit writing about Jamie and Stephen in hers, too, and Wendy in hers, at about the same time. Melinda's notes were fewer and ALS crowded everything stiffly out. ("The circus gig is turning out to be quite an ordeal, interfering with our ALS activities. . . .") I once asked Wendy if she would look through her notes and see if she could find anything at all. After a few days, she wrote back that she had found very little, besides the few, very happy diary entries she had sent me. "Sorry! During the time after Stephen's diagnosis I found it hard to write about my feelings. It was just too overwhelming to have to put it in writing, as if writing it down gave it some credence, like recorded history, and by not journaling what was happening I could maintain the possibility of miracles."

Matt During had told me over the phone that he had always been criticized for jumping around—and that now he had reached a stage where he could get away with it. So I was startled to meet a man who looked almost as young as Jamie.

"How old are you?" I asked.

He hesitated. "About forty-two," he said, and laughed.

We sat down in his office. "It's an interesting story," Matt said, in his strong New Zealand accent. "What motivates a lot of biomedical scientists is impersonal, emotionally distanced. It's unusual to take on a project because someone walks in the door wanting to save his brother." But he admired Jamie and his willingness to take chances, he said. "There, I'm somewhat similar."

I knew Jamie worried that Matt might be less than hopeful about his gene therapy for ALS. He sometimes thought Matt was more excited about his own brainchild, the neurovaccine. So I asked Matt what he thought about Jamie's plan. I expected him to say something cautiously optimistic.

"Most gene therapy to date has had a poor run," Matt said bluntly. First, there was too much hype. And everyone had been going after diseases where all else had failed, like inoperable cancer. Now they were going after easier targets. The brain made a good target, he said. It is hard for drugs to cross in, and it is hard to target just one part or another. With gene therapy they could select just the parts and cells they wanted. With so many neurological disorders, doctors have drugs that can help, but the drugs are not specific enough to be useful. Drugs can silence the parts of the brain that start epileptic seizures, for instance, but they silence so much else besides that they put the patient into a coma. A gene therapy injected into just the right part of the brain might do better. And neurological diseases often involve highly specific types of cells. In ALS, for instance, only motor neurons die, while the body's other neurons seem to stay healthy. Gene therapies, unlike most drug therapies, could be designed to target only the vulnerable nerve cells. "The gene is the most specific way to intervene," Matt said. And they were getting more sophisticated about switching genes on and off. They could inject a gene, switch it on to do its work, and switch it off when it had done enough. He predicted that gene therapy would have its first success in the next year or two—but not with ALS.

Next I asked Matt what he thought of Jamie's chances of treating Stephen within six months, while he was still in good shape.

"It's always hard to predict timetables," Matt said, just as bluntly. "Jamie's enthusiasm assumes we haven't hit any roadblocks. By December we will have some sense of whether we can move into a clinical trial. We need a controlled study." He wanted many animal tests before injecting DNA into Stephen's spine. Jamie was very hopeful, of course, Matt said. "How realistic that is—I'm not sure our first foray into this will be successful. I think we're a long way from understanding this disease, which always puts you on the bad foot. The disease is not simple. There's no agreement about the starting point." Naturally, because Stephen's disease was progressing rapidly, Jamie had "a sense of pushing," Matt said, but he thought their project would take longer than Jamie wanted. "Not as if I can go out and hire ten or twenty people. Certain techniques take time." He was not sure Jamie understood that. "He's trying to accelerate three to five years into one. But I'm not sure it will work."

That did not mean Jamie's plan was not worth trying, in his view. These experiments were necessary to advance the field. When parents of Canavan babies all over the world were pressuring Matt and Paola Leone to try again, during the controversy in New Zealand, he had defended himself in a letter to *The Lancet:* "Our concern for the suffering of the children helped to motivate this project, to offer some hope, not necessarily for these families, but for future generations of children who may have Canavan disease."

Matt During was Jamie's gene therapist and he was pessimistic about Jamie's gene therapy. Well, I thought, why am I so surprised? Why do I feel so burned? I know the odds. And yet I had begun to hope it would work.

Matt had been called a maverick for his Canavan work. "I have a reputation as something of a cowboy," he had said on the phone. But I could see that with this project, at least, Jamie was pushing the cowboy.

Matt wished his Canavan critics would remember the development of chemotherapy for cancer. "No one was cured overnight there. That took years of experiments with sophisticated combined protocols," he said. But with gene therapy there had been so much hype and expectation that people were primed now to react against it. "It's always politically a smart thing for scientists to say, yeah, we need to slow down," he said. "It helps your career because you sound balanced and mature. No one ever pats you on the back for doing things on the edge. They just call you a cowboy. It's ironic but sometimes the less you do the more you're commended for it."

I told Matt I was impressed by the size of his center, which was still expanding and—I could not help noticing—as disheveled as a new house on move-in day. He needed space, he explained, because he liked to change directions and charge toward the hot questions and the sudden frontiers. Here he spoke as grandly as Jamie. "I like to make a difference in as many aspects of medicine as I can. I always like to think of the provocative, the unexpected. I like to break dogma down. I have a lot of resources here at Jeff. Most of this research is not the sort of thing one can write grants on. You need something that is already far along before you can write a grant application. That kind of work is bread and butter. Ninety percent of science is bread and butter. Not very worthwhile. That's not what I want to do."

Finally I asked Matt to tell me about the second plan he was working on with Jamie, the neurovaccine.

Matt told me that he had just submitted a new draft of his paper to *Nature,* and he expected to hear back from the editors in about two weeks. He was racing a few other scientists toward neurovaccines. The month before, Elan, a company based in Dublin, Ireland, had published an account of an Alzheimer's vaccine in mice. *Nature* had run the story on the cover, because it was the first experiment that had ever slowed Alzheimer's disease—in mice. "Unlike complex procedures such as gene therapy," the editors wrote, "this immunization treatment is one that should be testable in human patients in the near future."

Jamie and Matt were developing both the gene therapy and the neurovaccine for Stephen. When he met Jamie, Matt had been working on vaccines for stroke and epilepsy. Now he wondered if it would help with ALS, too. If glutamate floods were doing the damage in all three diseases, then maybe his vaccine could make a difference. Matt wanted to test it. But he would have to reengineer the vaccine and target it to a different type of glutamate receptor, one that is important in ALS, a receptor called AMPA. If the vaccine worked against ALS, that would be great news, Matt said. It was powerful. "But we know it can cause changes in behavior. The rats learned mazes faster."

"What?"

"We did a crazy experiment here in Philadelphia," he said. He told me about the vaccine that he and his team had made to help protect against brain damage after a stroke. They made the vaccine and gave it to rats. Matt ripped down a chart labeled "Figure One" from his bulletin board and showed it to me. "Multiple injections," he said, making the gestures of an addict shooting syringes into the crook of his arm. Then he and his team had induced strokes in the rats. The injections had prevented seizures in more than seventy percent of the rats, and those that did suffer seizures were protected from brain-tissue damage.

"An interesting story," Matt said. "But like anything in biology, it gets more complicated. We did a screen for behavior. The rats were actually *smarter,*" he said again. In maze studies the neurovaccine had had the same sort of effect on the rats as his gene therapy for memory had done.

"It's pretty novel," he said, with a smothered smile of satisfaction.

The neurovaccine had many other innovative features, Matt went on. It broke ground at every step. The sheer amount of novelty made it revolutionary, in fact, which is why the editors at *Nature* had already rejected it once; but revolutionary research was the only kind that interested him anymore. Let other people worry about old stale dogma. Neurovaccines would transform medicine in the new century. They would be used to attack the plaques of Alzheimer's disease and help the brain heal itself of prion diseases.

"And Lewy bodies?" I asked.

"Sure," Matt said.

How hard it was to speak up and identify myself as not only a reporter but a family member in need. I felt hope, dread, and shame. Speaking meant dragging myself up and over a stone wall. I tried to keep my voice under control. Thinking about that moment later, I understood better why Jamie had hesitated to tell his news to his office mate and to the director of the Neurosciences Institute. For me this conversation was the beginning of a whole new appreciation for what Jamie was about. What a world of difference there is between asking questions as a professional and professing a desperate need. You stop being a doctor or an engineer, an entrepreneur, or a reporter. Suddenly you are just a human being who is very frightened and trying not to beg.

"The reason I asked you about Lewy body," I said, "is that my, ah, mother has that, unfortunately."

"Oh, she does? I'm sorry."

"We're watching a rather rapid decline right now," I said. "Of course—" I went on, trying to keep my voice calm, but finding myself trapped in sentences that were strangely hard to finish, "of course, it's very different from watching your kid brother—my mother is in her mid-seventies—but still, you wonder whether something like this—"

"Well, yeah," Matt said. He looked at me. He had been through many conversations like this one. "It certainly could be done," he said. "I mean, if I had the resources, we could do something that would be worth considering there. One of the key proteins in the Lewy bodies is alpha-synuclein. Alpha-synuclein is found in all sorts of amyloid deposits, including Alzheimer's and Parkinson's, and, from my understanding, Lewy bodies. And that would be a good protein to target with the immune system, I think. If I had the resources, I could do something."

I had stopped making notes. I let my tape recorder roll while I watched Matt sit with his fingers to his chin and read the screen of the computer on his desk. He was engrossed in the same kind of PubMed search that I had often heard Jamie Heywood make when we talked late at night on the phone. Matt, too, seemed to have a way of losing himself in the monitor.

"Hey, this is interesting," Matt said, and fell silent again. I waited, listening to the clicks of his mouse and the clicks of the computer. If I had taken two steps I could have looked over his shoulder, but it did not occur to me to move.

"The key is, if they have a model," Matt said, half to himself. To test a vaccine, he would need a Lewy body mouse—the equivalent of the ALS mouse.

Matt glanced up. He would need ten thousand dollars from me for the animal tests, he said.

My eyes must have widened.

"It sounds a lot, but it isn't, really! It would be consistent with what we're doing," he went on, judiciously. "I would be happy to get involved. The research is straightforward. The red tape is harder. But alpha-synuclein may be a good target for Lewy body, and Alzheimer's, and Parkinson's. We would target the immune system to the plaques, and clear the debris."

Matt was telling me why he would to do this for me. Alpha-synuclein would be a huge prize. The very same vaccine that he developed against Lewy body dementia might work against Alzheimer's and Parkinson's. I was offering him another ship of opportunity.

He told me that he could get special permission from the FDA, the kind of emergency permission called "compassionate use," in order to immunize my mother against Lewy bodies, just as he would get compassionate use to try an ALS neurovaccine on Stephen. The bureaucracy makes allowances for special cases in which individuals need to move quickly: Out of compassion for desperate people in medical crises, the red tape is cut to a minimum. First Matt would need a brain biopsy for confirmation of the diagnosis, he said, staring at the screen. Then he and his team would genetically engineer a vaccine. The vaccine might induce my mother's immune system to clear her brain of the deposits that had progressively impaired it. Matt and Paola could get rapid approval from the FDA on the basis of Elan's Alzheimer's vaccine, which the FDA had already approved for a safety trial. Immunization might help clear up the Lewy bodies. As we both knew, Matt went on, there was no other treatment, and the disease was progressive and fatal.

"If it was my mother, I'd be doing that. It's certainly worth doing. We could move quickly."

I told Matt the names of my mother's neurologists. There was Stephen Salloway, and there was the man she had seen at Harvard—a doctor with a name like Seltoe, I said. Why could I not remember that name? It was a peculiar experience pronouncing these names from our family's secret saga—as if I were dragging them with me over the stone wall, too.

"*Selkoe* is very famous," Matt said, suppressing a smile, like a psychoanalyst whose patient has just mispronounced the name of Sigmund Freud. "He's a world expert on Alzheimer's. He may be too established to get." Of the two doctors, Matt thought Salloway was the better choice for our purposes. I should call him and ask if he would sponsor a trial of the vaccine. I should tell him that Matt would consider generating the vaccine. We would need Salloway to sponsor it as my mother's doctor. There would be a reasonable amount of work from Salloway's side, Matt said. "The procedure is quite involved, unfortunately. One needs a gift for politics, too, to ease the wheels. This would be only one case. We couldn't use it for publication." So we would need a physician who was motivated, like Salloway. "If it hits a roadblock, Selkoe may be overextended."

On reflection, Matt said, we would not want to wait for an animal model. We would do a toxicology study. I might have to put up ten thousand dollars for that instead.

Suddenly it was Matt who sounded nervous. He said he was worried about telling competitors what he was doing with his neurovaccines. If I told anyone that he was working on vaccines to the brain, I might jeopardize the paper he had submitted to *Nature.* He was more nervous about Selkoe than Salloway. (I realized why later. Selkoe was involved with Matt's competitors' work on an Alzheimer's vaccine at Elan.) On second thought Matt preferred that I say nothing about the work even to Salloway. I should ask Salloway to call him directly.

We stood up, and Matt gave me a tour of his vast maze of a lab. I was preoccupied. I had never felt so hypnotized and so distracted. Though I had my reporter's notebook in my hand, and kept scribbling away out of habit and reflex, my imagination kept leaping ahead to my mother, my father, and my brother. It was remarkable how that brief exchange with Matt in his office had transformed the look of the situation for me. There was a chance that my work as a science writer could save my mother and clear up whatever had fouled her inner weather for the last dozen years. What would my family say? Would we try it? Would it work?

Matt led me on through the hodgepodge maze of rooms. On a laboratory bench, a white rat lay in a pool of light, attached to a micro-injection pump. The rat was lying on the stage of a stereotactic apparatus: a dissecting scope with binocular eyepieces. Two young scientists had drilled holes in its skull. Now they were pumping in DNA. As I looked at the rat, I did not picture Stephen, or my mother, lying there on the table, because the thought was unimaginable, but I could not help staring at that rat. It was alive. Its whiskers were twitching.

In another room, Matt showed me where he and his postdocs were growing bacteria and manufacturing DNA for his gene therapy projects. A young postdoc by the vats introduced himself and told me that he was getting over a mysterious flu. He was just back from New Zealand. He gave me a wet handshake.

"Don't worry," he said. "Just washed my hands."

In still another room, Matt showed me a cage full of mice that Jeff Rothstein had sent up from Johns Hopkins. Then Matt introduced me to another postdoc, Dave Poulsen, from Montana. Dave was using a gene gun to inject genes into mice. He put it in my hand. It looked and felt like a plastic ray gun, made of computer-case gray plastic. The spent shells on the laboratory bench were plastic, too. The gun fired tiny gold particles, one micron in diameter coated with DNA. The gun shot the gold into the skin of a mouse, a rat, or a human patient with a burst of helium gas. The gun was armed and ready to fire. Dave demonstrated the procedure. The gun produced a high-pitched beep as it fired into the rat's skin. It was nowhere near as painful as getting a shot, Poulsen said, but he anesthetized the animals anyway.

Poulsen had shaved the bellies of the rats before the injections. He demonstrated again and I could see a hint of gold where the shot went in: a faint gold spot that looked enormous to me but was about the size of a dime.

This was the postdoc Matt had assigned to handle the preparations for Jamie's gene therapy. He was in his mid-thirties, a bit stout, with a red-and-white-striped polyester shirt, brown hair, a beard almost as short as stubble, and the kind of aviator glasses that were in fashion back in the 1980s. The postdocs I had just met all looked as if they felt they were on a vertical expedition together, as if they were all hands on a climbing rope. Dave Poulsen looked like a man who would rather be someplace else: A decent guy, but even with a gene gun in his hand, he was not a bolt from the future.

"I have a lot of rats here," Matt During told me, with a little laugh, as we went on with the tour. A few of them were pets from the memory-enhancing gene therapy and the memory-enhancing neurovaccine. "I kept Jeremy. He has a 190 IQ."

Walking back from Jefferson Medical College to the train station at Market East, I was amazed at the way the names of my mother's doctors, Salloway and Selkoe, were transforming themselves. So were the words of the diagnosis, Lewy body dementia. Up until that moment, the names and the terms had all been gray and heavy as lead, synonyms for despair. Now that there might be something we could do, it was almost as if the lead was turning to gold. There is something anesthetic about resignation, and I felt suddenly as if a painkiller and a tranquilizer were wearing off. After that dark talk with my mother by the coatrack at the foot of the back stairs—*Hold on*—maybe I had found help after all.

At the train station I called Deb from a pay phone to tell her that I was catching the 1:47. Then I could not resist telling her what Matt During had said, although as I talked I realized how wild it all sounded.

"A brain vaccine!" she said.

The more I explained, the wilder it sounded to me, too.

"Don't miss your train," she said.

In a seat by a window I pulled Matt's stack of papers from my briefcase, and found the manuscript of his unpublished neurovaccine paper. I underlined the phrase "neurodegenerative disorders" in the abstract. But after one page, I put it down. Again I felt the weight of what Jamie was doing for Stephen.

In the window of the train, the leaves of trees rushed by backward. A single sparrow hopped in a bush. I stuffed Matt's papers back in my briefcase. What I was feeling at that moment was what

Stephen and Jamie had felt when Jamie set up shop outside his Cave: the collision of resignation and defiance. After a while it is almost comfortable to let an illness progress, even a fatal illness, if you are absolutely sure there is nothing that anyone can do. And in that dull anesthetic comfort it can be painful to let yourself hope again. Stephen had lost some of that comfort when he decided to support Jamie in his fight. He had to give up something. Each brother had given a gift to the other.

Again I thought how level Jamie's head was that first evening, the evening of the diagnosis, back at the Neurosciences Institute. My emotions on the train were only to be expected. But Jamie's level head and his ability to decide and command—that was something rare.

My eyes felt tired. I did not want to rub them, because I had not washed my hands since the lab. Walking with Matt During in a borrowed white lab coat past all those vats had not bothered me: I was used to laboratories. But now I stared out the window and worried about that wet handshake. Well, it was a distraction from what really worried me.

{ TWENTY-FOUR }

Thunder and Lightning

"Put the right people in a room, stir, let simmer, bring to a boil," Jamie exulted to me on his cell phone from the car as he left one of his lab meetings at Matt During's. It was mid-August 1999. Jamie was back from Paris and he was doing a lot of that—calling me from his cell phone to check in and whoop at how well things were going. He and Matt were meeting with scientists and the vice presidents and presidents of biotechs. "I swear to God!" he exulted after a lunch in Philadelphia. "This is the cheapest way to buy people's time there is. You get five people talking about ALS for two hundred bucks? It's a deal!" After one dinner in New York, he told me that he had just dropped $900. "Not holding back. Which is actually my philosophy on how to do everything."

He was very happy with Matt and Paola. "These are aggressive people. They're not . . ." He hesitated—I think he was trying not to put down the pure scientists I usually write about. "I think it's so easy in science to *not do it.* Matt really wants to make science happen in the real world. I think he just cares about that a lot." I heard what Jamie was not saying, and felt indignant and annoyed on behalf of scientists. Without them, Jamie would not have his EAAT2 project. Without people like Jamie, Matt, and Paola, it would not be happening either—not this soon and this fast. On the other hand, scientists were telling me almost unanimously that this EAAT2 project was too soon and too fast. When Jamie put down scientists, I felt as exasperated with him as he was with scientists. I challenged him on this once, later on. Jamie said, "To say scientists are slow or lazy is not accurate. Most of them work very hard. But in modern medicine, scientists are reluctant to jump from researching in yeast or mice or rats to trying to treat humans. In desperate diseases, they should be able to move quickly from the lab to patients. They could do that fifty years ago—we were much freer back then. But I think medicine today has a culture that is overprotective and makes it very difficult for researchers to make those leaps. I think their caution is hurting us as a society and we're not moving forward as we should. There are complex reasons for this, and some are valid, but I think as a whole we're just way too reluctant to treat people with new ideas. Somehow too many scientists lose themselves in the question of what goes wrong in ALS and lose sight of the fact that twenty people die of it in this country every day."

Jamie also called me at night from his desk in the Victorian. He surfed the Web for ALS news while he talked. With the Web, biomedical insiders and even outsiders could now monitor global research instantaneously, which even the greatest medical specialists could not do a decade before. So Jamie would scan PubMed while he talked to me, and sometimes I would hear his rapid-fire voice begin to slow down. "Oh, look at this . . . Holy shit . . . I've got to e-mail this to Matt."

"What is it?"

"Well, this study, which came out . . . in Poland . . . *two days ago . . .*"

As Ben said, it was the ALS channel: all ALS, all the time. Jamie's personal life was getting squeezed out to the edges. Melinda was pregnant, and even that good news was cast in the strange, harsh light of ALS. That night, just before he hung up, Jamie said, "Oh, we went to our sonogram today. We saw the baby. The nurse was looking for the facial shot, right? And then we kept getting this great shot of this curved spine and the skull." Jamie laughed. That was the part of the human anatomy he had been studying compulsively for the past half year. That was the model that sat on the desks of almost all the ALS experts he had visited. He had been learning his way around images like the one in the sonogram ever since the night that Stephen had called him after his appointment at Massachusetts General.

"You know," Jamie said, "that was the baby picture I wanted, and it was the only one I didn't get, because the nurse didn't think that anyone would want to look at a curved spine and a skull. But that was the photograph I understood."

The baby was due on April 20, 2000. If he stayed on schedule, by that time he should have genetically reengineered the nerves in Stephen's spine.

The more Jamie learned about gene therapy, the more he wondered if the nonprofit route was the way to go. Jamie was beginning to think if his gene therapy project should be for profit after all. He began talking of upping the ante again.

Not everything in the gene therapy world was nonprofit. Most of the work was for profit. According to a study by the NIH, more than half of the gene therapy trials registered with the RAC in 1999 and 2000 were paid for by private money. That was true of James Wilson, across town from Matt During at the University of Pennsylvania. Like Matt During, Wilson was a rising star in the field of molecular medicine. He had been thirty-seven when he joined Penn as the John Herr Musser Chair in Research Medicine; and director of Penn's new Institute for Human Gene Therapy; and professor at the Wistar Institute nearby in West Philadelphia. His institute was housed in a new sixty-million-dollar facility at Penn's medical school.

Wilson had founded a biotech company, Genovo, based in Sharon Hill, Pennsylvania, four years before. He owned nonvoting stock in the company and the company paid Wilson's lab within the institute almost five million dollars a year.

Among many other projects, Wilson was working that year on a gene therapy for ornithine transcarbamylase (OTC) deficiency, a rare metabolic disorder. Babies born with it—mostly boys—cannot break down ammonia efficiently, and most of them die soon after they are born. Normally the liver breaks down ammonia so the body can get rid of it in the urine. As the ammonia builds up, the baby goes into a coma and dies.

Volunteers in most gene therapy trials are, like the parents of Canavan babies or like the family of Stephen Heywood, looking at a situation so desperate that they are willing to take almost any risks. But Wilson's OTC trial was unusual for the field because he and his team included people who had milder forms of the disease. All volunteers, healthy or dying, know that they may lose their lives, and that the benefit may only be for people with their disease years afterward. But most healthy volunteers have many years to lose.

That spring while Jamie and Matt raced to prepare a gene therapy for Stephen, a young man named Jesse Gelsinger had enrolled in Wilson's trial. He had a treatable form of OTC and could survive with the help of pills—he took about fifty a day—and a strict nonprotein diet. The gene therapy was not designed for people with a treatable case of OTC. It was meant for newborn babies who would die without it. Wilson and his group thought the gene therapy they were testing was safe enough to include Gelsinger even under those circumstances.

The boy's father, Paul Gelsinger, a handyman in Tucson, Arizona, had raised him; Jesse's mother was more or less out of the picture. Paul read the eleven-page patient consent form and cautioned his son. He made what he felt was a reasonable guess and said, *There's risks here. There's a one in five thousand chance of complications.*

The boy looked at his father. *I can beat that, Dad.*

His father took a snapshot of him on his eighteenth birthday, which he chose to spend enrolling in the trial. Later that picture appeared around the world: the young man posing in front of the famous statue in Philadelphia of Rocky, the underdog prizefighter, Jesse Gelsinger flexing his biceps with his fists raised to the sky. It was the first big thing he had done in his life.

Matt During and Jamie Heywood decided to form a biotech company together, though they were still not sure if their EAAT2 gene therapy should be profit or nonprofit. Meanwhile, in the welter of projects that Matt was running or proposing to run, Jamie kept working with all his charm, energy, and cunning to keep the gene therapy moving fast. He was in a delicate position. He was the project manager and he was also an outsider at all of these labs. And both Matt and Jamie were continually flying into new projects. Now that they were thinking of starting a company together, there was, besides the urgency of saving Stephen, the urgency of getting ahead of the competition. With their jockeying with each other, and their jockeying for their projects, and their jockeying against the clock and the competition, meetings between Matt and Jamie began to get complicated. Jamie still had one goal: to save Stephen. But Matt wanted to advance the molecular revolution all along the edge of medicine. At every meeting Jamie would try to focus Matt's attention on the gene therapy while Matt would pitch Jamie six new dreams of revolutionary therapies, which he thought they should add to their new company's growing list of hypothetical projects. They were two cowboys galloping together.

Jamie would groan. "I'm getting burned."

"I've got a few more ideas."

"If you give me any more ideas I'm going to pass out."

"Let's see, we've got the botulinum toxin vaccine, which may or may not be a good idea. . . ."

"You're reaching on that one, Matt."

Because Jamie wanted Matt to move as fast as possible with the gene therapy, he wished Matt had not assigned the first phase of the work, the development of the DNA package, to Dave Poulsen. Poulsen was the postdoc who had shown me the gene gun and the little golden spots the size of dimes. It was his job to engineer the virus that would carry the DNA into the spines of the hundred test mice and then into Stephen's spine. Poulsen was working at this task very slowly and methodically, at a pace that drove Jamie crazy. He wondered if Matt would have given his gene therapy project to Poulsen if Matt had believed it might save Stephen.

Early one morning in September, Jamie dropped by Matt During's office as if he lived around the block, and had not driven down that morning before dawn from Newton. "Just stopped by for guidance," Jamie said. "Without taking more than two minutes of your time."

Matt did not seem to hear him. He was reading a paper about stem cells on his computer screen, with his usual absolute absorption. After a moment, he looked up and offered to print out a copy for Jamie.

Jamie nodded.

"The press release or the paper?" Matt asked, poker-faced. That was a dig. Press releases are issued by journals for science writers and other outsiders who may not be able to follow their articles. "I'm sorry," Matt said. "I'm being insulting, I know."

Of course, Jamie really was out of his depth most of the time, and by now Matt knew it. On the other hand, Matt himself, in his fever to start revolutions all along the edge of medicine, could get out of his depth, too; and Jamie knew it.

"I'm sorry," Jamie replied in the same brotherly tone. "Are *you* an expert on stem cells?"

"I've been to—" Matt rattled off a list of lectures and international seminars. "I know a little bit about stem cells."

Like Dolly, stem cells were a bolt from the future, and in the fall of 1999 the future was still looking like an Eden re-made. Dolly herself had given birth to a lamb, and her Scottish vets called her Bonnie: "a bonnie wee lamb." With stem cells, biologists could now hope to build and rebuild living bodies the way architects built and rebuilt cathedrals. They could dream of taking any cell from the body and coaxing it to become the cornerstone of a new cathedral. They could govern the growth of life and they could shape the repair.

Matt turned at last from the screen and faced Jamie. He had just gone to an informal talk at Jefferson about stem cells, he said, and now he had an idea for a stem cell project that might help Stephen. The results he had heard at the talk were not even published yet. But if Matt put them together with this paper on his monitor, they suggested that one might inject stem cells that normally make blood, hematopoetic stem cells, into a brain, and the cells might make new neurons instead of blood.

Matt handed Jamie a copy of the new paper in *Science* about turning brain into blood. "Which is the opposite of what *we* want to do. But the same principle applies."

"Unless it doesn't," said Jamie. Matt's idea sounded wild to him. In any case, Jamie had decided to let other people in the ALS community work on stem cells. He wanted to stay focused on his gene therapy. A visit with Matt During was like a visit to the alchemist, the wizard, the wonderful wizard who keeps dancing off into a dozen fairy-tale projects and possibilities.

Jamie did not want to go near this with his ALS Therapy Development Foundation. But he could consider it for their new start-up, a little biotech company they called CNScience—Central Nervous System Science. And if he and Matt kept it entirely separate from the foundation, then there would be no ethical conflict with his making money from the stem cell project.

Matt pitched. Jamie sat in his chair in the position of the Thinker, leaning toward Matt. "Do you have people to do this experiment? What's the name of the guy at Jeff?"

But Matt did not answer. He was staring at the monitor again.

"What's the name of the guy at Jeff?" Jamie asked again.

"Yeah, it's pretty interesting, actually," Matt said, as if he had not heard. "People who work on blood don't think about the brain. You need to remember that. They think about cancer." He had thought of this idea only because he knew a little about stem cells and he knew a lot about nerve-death diseases.

"Will anybody else look at this and make the connection?"

"Yeah. People like Mulligan. So we've got to move quickly. It's a big breakthrough."

"If it works. OK, we're off-mission here, though. I'm monopolizing your time."

But Matt ignored him again. Just the day before, Paola Leone had gotten a shipment of human stem cells for their Canavan work. Now Matt saw an opportunity not only for treating Stephen but also for their new biotech venture. Matt leaned toward Jamie and explained what he had in mind in a high-velocity, slightly conspiratorial voice like Jamie's own. "The key, from the intellectual property point of view, is *just to do it,*" Matt concluded. "I don't think anyone else is going to move as quickly."

"The ALS community has pretty much accepted that stem cells are nonharmful," Jamie allowed.

"This is something really worth doing quickly. You could blow people away. You could just blow people away."

"OK. I'll write this up, too," Jamie said, in an executive voice. He made a few notes, looking concentrated and decisive. He would add stem cells to their list of candidate projects at CNScience. I think the chance of getting out ahead of everyone else tempted him almost as much as it tempted Matt.

"Nothing's been published yet," Matt said again.

"The only people who know about this were at the meeting," Jamie conceded. "Small group."

"But there were high-powered people in that group. High-powered. So we need to move quickly."

Next, Jamie dropped in on Paola Leone in her office down the hall from Matt's. With Paola he knew he could focus on gene therapy. Paola played the same role in the Canavan gene therapy project as Jamie played in his gene therapy project. She was the manager; she was there to usher the project along. On the walls of her office, she had hung some of the photographs she had made of toddlers with

Canavan. One was a beautiful picture of babies and toddlers propped on a couch, helter-skelter, like dolls. They were resting on the couch, and somehow because none of them could move very much, they looked as if they were floating in space in zero gravity. All of them were tiny, with heads too big for their bodies, and they were lying just as they had been set down. Some Canavan babies were able to function longer than others. Propped on a shelf, Paola had a framed drawing of a frog, scribbled in green crayon, and inscribed by a parent, "To Doctor Paola, Love Max."

The calendar on the wall was international. It said "September *Septembre Septiembre* September." Her windows looked out across a narrow courtyard at the yellow brick walls of the medical building.

"I have to download you," Jamie said briskly. He wanted to know all about her experience with Canavan, getting experimental gene therapy approved by the FDA.

Paola described the details of the approval process in the United States and in New Zealand.

"That's kind of all off, now, isn't it?" Jamie said. "Running trials in New Zealand is harder."

"Yes, because of this," Paola said tersely, nodding toward the photographs of all the babies they had worked with in their Canavan trial. They would not be able to do that again in New Zealand anytime soon. Now it was up to the FDA whether Jamie's gene therapy and the Canavan project would go ahead. And Paola pointed out that there was one consideration in Jamie's study that they had not had to worry about with the toddlers. "Because your ALS patients are sexually active, they'll worry about the germ line." The FDA would want to see animal safety studies, including primates, to make sure the DNA injection did not find its way somehow into eggs and sperm.

Paola's phone rang. It was Matt, asking for stamps.

"Stamps?" Jamie cried, after she hung up. "Doesn't his secretary handle that?"

"I'm going to teach you about dynamic incompetence," Paola said cryptically, and laughed. "This is the way I make him dependent."

Jamie ran through his list of projects with Paola and explained why he was most interested in the gene therapy. "I have patients begging me for help right now," Jamie said. There were risks with gene therapy but they did not worry him as much as some of the other projects on his list. "The neurovaccine scares me," Jamie said. "That could rip out every nerve in your body."

Paola explained the ins and outs of FDA approval for gene therapy while Jamie made a few spare notes in his white legal pad—half a dozen bottom lines, always bottom lines. His conversation, too, was always bottom line now: He spoke not impatiently but briskly, the model of a project manager. He flipped through Paola's draft of her Canavan proposal to the FDA. When she handed him the paperwork that she was preparing for their next Canavan trial, Jamie skimmed through it, stared, and gave a small smile. Paola had just handed him exactly what he needed. The Canavan project was so much like his own that he could imagine modeling his on hers; it might save him months. Suddenly he blushed, bowed his head as if he did not want Paola to see his face, and ran a hand through his hair.

"It's the same as mine," he said. "Just a different gene."

"Exactly."

Jamie told Paola that he would be testing the gene therapy on one hundred mice.

"One hundred mice, that's huge."

"I know."

"If something goes wrong with just one, a stroke or something, there may be more questions—did that experiment cause that problem?"

"I ask the question, what is going to make me confident enough to treat Stephen."

"Hmm."

With the brain vaccine, Jamie thought he wanted one or two people to try it before Stephen. "The question is, what are the rules on what you can do? Does that have to go to the FDA? If you want to try it on late-stage patients, I can get patients that are months from death, guaranteed. There are people who are contemplating suicide at the same time as they are contemplating something radical."

"I think it can be done if you have a very clear consent form," said Paola. "But do you really want to do that?" In the Canavan trials, she had had parents begging her for two years to inject DNA into their babies' brains. She had never forgotten how some of those parents reacted when their baby had a seizure. "They are looking at me like—"

"You're screwed."

"There is a huge gap between what they *say* and what they will *do*. Maybe you should take this overseas."

But Jamie did not want to do that. If the gene therapy or the vaccine looked to be effective, he had to try them soon and in such a way that he could offer them to the many people who needed them. Otherwise everyone alive who had ALS would be dead. "I don't want to do this on the sly. I want to do this right. I don't want to bend the rules. Because it's important."

"Why don't we just look at this from a very simple perspective?" Paola said. "You don't want to do harm. So why don't you focus on toxicity? Just make sure that what you want to do is not going to harm patients."

As they talked, Paola doodled in her own legal pad, drawing simple brackets and parentheses. Her page was even more spare and abstract than Jamie's, with no words at all. She was such a passionate scientist that she saw the work almost as impatiently as Jamie. When they gossiped about a new arrival in another lab, Paola told him, "He's a scientist, you won't like him." And they both snorted.

Now Jamie brought up the tricky subject of the postdoc Dave Poulsen. He explained to Paola that Matt During had given Poulsen the key job of making the construct for the gene therapy test in the mice, and Poulsen was holding things up. In Baltimore, Jeff Rothstein was ready to go—he had the mice and he was ready to inject them with the DNA. Poulsen had had the stuff for fifteen days now and he just sat on it.

Paola closed the door, and confided that she was not fond of Dave Poulsen. She had problems with his work. He was their slowest postdoc, she said. He never worked on weekends, never stayed at the lab past two or three in the afternoon. "Which is great. We should all live that way."

"The mice are ready to go at Johns Hopkins," Jamie said again. "They're ready to go. It's an expensive trial. We need to get them treated. We've got seven mouse trials waiting at Hopkins! Well, I can get Dave excited. I'm good at this. I'll fly him up to Boston to meet my brother. I'll give him a consulting fee, $5,000."

"That'll help. Money talks."

Jamie treated Matt and Paola to lunch, an expensive lunch, and more than once he steered the conversation to Dave Poulsen. Matt listened like a politician; Paola looked upset. The more she listened, the madder she looked. She told Matt that they had to try to light a fire under Poulsen or fire him.

Matt listened with a slight smile. He told me later that he thought Poulsen was working well. Jamie and Paola were being too hard on one poor postdoc.

"He's killed other projects!" Paola cried. "We've given him millions of other chances!"

"Hey!" Jamie waved his hands. "End of topic! End of topic! OK? Here's what we need to do." He laid out a strategy for the lab meeting that afternoon. "I mean, no mistakes. We don't have time to make mistakes."

After lunch, Paola Leone, Matt During, Jamie Heywood, and half a dozen of the lab's postdocs pulled up chairs in Matt's office. The calendar on the wall was from the Louvre. The photograph from September was a pair of marble hands sculpted by Rodin to form something like the spire of a cathedral. A white lab coat hung on Matt During's door. A candle flickered on the windowsill in the hot city air of Indian Summer.

The postdocs draped themselves comfortably around Matt's office. Like him, most of them chatted in broad New Zealand accents. As usual they had the air of cool, droll partners on a climbing rope, and so did Matt. There was always some of that feeling in Matt's office: the planning of a difficult ascent. But Dave Poulsen sat against the far wall from Matt, looking stolid. Paola leaned against the door-jamb and glowered at him.

Long afterward, I asked Poulsen about that day. I tracked him down at a lab in Missoula, Montana. "There were a lot of experiments needed to be done," he told me over the phone, "and to get them done in time I had to live in that lab twenty-four/seven. That's partly why I left and came back here. I had a family with three kids. I'd leave home it was dark, I'd come home it was dark. Weeks would go by when I didn't see my lawn, let alone my kids. These experiments aren't things that happen overnight. Science takes time. Jamie wanted us to give him something he could put into his brother tomorrow. And he felt like anything short of that was just—being lazy, or that we just simply didn't understand. And I think it was just the other way around. Jamie just didn't understand how science worked."

That afternoon, Matt and his group attacked the chief engineering problem ahead of them: designing a DNA package small enough to fit inside the virus case. Jamie was right in there with them. They were trying to figure out how to leave out 640 base pairs.

Now and then Jamie ran his hand through his hair and leaned back in his chair: "Ahh, you're right, you're right!

"It's packageable but it's big," he said at last.

"It's a little optimistic but I think we can do it," said Matt. "Dave?"

"Well, we've got a couch out in the hall now," Dave said gloomily, in the voice of Eeyore, "so I'll have a place to sleep."

"I'm behind the eight ball with Jeff Rothstein," Jamie said, speaking not to Dave Poulsen but to Matt During. "He had expected to start in August and here it is September." He was asking Dave Poulsen to move—but he was asking delicately.

Outside, there was a murmur of thunder.

Jamie shifted topics and talked about places to inject the stuff into Stephen's spine. They all joined in that conversation, and as they did they pointed to their own backs. "Cervical is . . . ?" Jamie asked, and pointed. "Yeah," said Matt, and pointed at a spot on the back of his own neck. As he did there was more thunder.

This was one of the problems that scared Jeff Rothstein in Baltimore. Matt and Jamie hoped the DNA injection would not only improve strength in Stephen's arms but might reach the bulbar region, which would improve breathing and speaking. "I don't think so," Rothstein said, when I asked him about that. "I know that's what they hope. I don't think so. The bulbar musculature is controlled by a part of your brain called the brain stem. And that's within the cavity of your cranium, which is actually above the top of the spinal cord. So I doubt it. I guess it's theoretically conceivable that their injection of fluid would actually diffuse far enough. I doubt it.

"And the flip side is, if it migrated that far, it could also be toxic. And that's part of the brain you do not want to poison. It's that part of your brain that regulates all of your core breathing functions. That's where people shoot themselves in the head and kill themselves."

There is a legend that Cain did not know how to kill Abel. Out in the field, he worked his way up Abel's body until he found the throat. Jamie was his brother's keeper, but Jamie did not know how to save Stephen, and now here he was working his way up to the back of Stephen's neck.

At just that moment, while Jamie and Matt were arguing about where in Stephen's neck to insert the needle, Dave Poulsen interrupted the meeting. "I've got to get home early today," he said. "Is there anything else we need to discuss?"

Lightning and thunder. "Thought it was my ears ringing but it's not," said Matt, distractedly.

Dave Poulsen stood up. He gave them all a big, grizzled, mournful grin. He missed Montana, he said. "Montana was therapeutic.

This is the opposite of therapeutic." At the door he paused and said again, "So I guess I'll be sleeping in the lab."

Jamie stood up, too, and shook Poulsen's hand. He gave him one of his small Superman half-smiles. "I'll buy you a pillow. An ALS pillow. You can call me in the middle of the night. I'm always up."

Part Four

The Sign

To do science, you must never lose a child's hope.

Günter Blobel

TWENTY-FIVE

Jamie's Old Room

In late September I booked a flight to Boston. On my way to the airport I stopped in Princeton to teach a seminar, "Science and Literature: Parallel Lines." It was the second year I had taught the course. The year before, I had focused on the young writers around the conference table, and the course had gone well. But this class was not beginning well. I could not stop talking about the Heywoods.

Jamie had offered to put me up for a few days at his parents' house on Mill Street. He and Melinda were not ready to put people up in the Victorian, he said. They were just squatting there and most of the rooms were still empty, bare and raw. I would be more comfortable at his parents'. Or, if I needed my own space, I could stay at the Newton Hilton.

Normally it is better for the mental health of writers and subjects if they spend some time apart during long days of interviewing. Otherwise profilers and their profilees get jumbled together in each other's heads like cubist portraits. But I wanted to see the Heywoods at home, so I told Jamie that I would accept his parents' offer and stay with them. I felt awkward and a little ashamed at the thought of staying with them in the middle of their crisis, even though the Heywoods wanted the publicity.

In retrospect, I can see that the story already was jumbled in my head. The closer we came to the moment of truth, the more confusing the picture looked. I could no longer tell how much of our hope was real; and sometimes, when he was at his most electric and charismatic, I still wondered how much of Jamie was real. That past spring, in the Caribbean, Jacques Cohen's babies had looked like a Solomonic problem. Now that we were in the thick of Jamie's race, it looked monstrously complicated, too. The stakes felt higher every day. I bounced around among them like a pinball.

At the time I felt ashamed of my own confusion. But confusion was inherent in the project. None of us could see far enough ahead to know what was right to do. Our hope and fear seemed to grow, the closer we came to the crisis. We were all agitated and we were all looking for signs.

Jamie met me at Logan Airport. As we drove along Memorial Drive, on the Cambridge side of the Charles River, I glanced across to the Boston side and saw an amazing building gleaming in the dark above the water. It looked like a morph between a grand New England Victorian and a millennial palace, lit dramatically to give an impression of dark brick wings and much glass. It was a beautiful sight, ghostly and brown, Boston transfigured into a dream of the future, floating by itself above the dark river.

"Genzyme," Jamie said. He told me that it was a biotech company dedicated to orphan diseases. "They built the whole company on one disease they've cured," he said. "Maybe one of the Hassidic disorders. A protein therapy. They make about 130 million a year off it."

I looked this up later. Genzyme's first drug, Ceredase, was a treatment for Gaucher's disease, a rare genetic disorder that can cause an enlarged spleen and liver, and neurological problems. When the company won government approval to market Ceredase in 1991, people at Genzyme thought only about one thousand people would use it. But the incidence of the disease had risen. That year there were nearly 3,000 people using Ceredase around the world. Under the Orphan Drug Act, a company that gets FDA approval for a drug first is granted seven years' exclusive rights to sell it in the United States.

Genzyme charged patients or their insurance companies as much as $170,000 a year for Ceredase, so the drug was a bonanza for the company. That year Genzyme was expecting sales of almost half a billion dollars.

"ALS could represent a fifty-million-dollar market," Jamie said as we drove on along the river. "People will fight for that." Jamie was wrestling daily now with his choice of nonprofit or profit. He admitted that he still had the start-up itch. He was sure his gene therapy would be huge if it worked—and getting big-money investors would be a much faster way to get a lot more money for the project. Which way could he develop his therapy fastest? Should he partner with a company like Genzyme—fold himself into someone else's company and lose control? Or should he retain control and build a start-up of his own? Running a nonprofit might be three or four times cheaper, by his reckoning, because people trust nonprofits. When no one stands to get rich, the surgeon volunteers. Hospital costs stay low. The FDA plays by different rules. If you do not have anything to gain, people believe in your data and intentions. Of course, if you sell your idea to Genzyme, you have a powerful company to make and market your drug. But the Heywoods' friends and family and the congregants at Grace Church were not giving Jamie money so that he could get rich, and that is what would happen if he sold his gene therapy to Genzyme. "This is a really difficult topic," Jamie said. "At some level, some little old lady's dollars are going to make somebody money."

In the dark of the car, I looked at him. Occasionally I remembered that Jamie's chances of success were still very small. More often, I felt as if he might be just a turn of the road from building something even grander than Genzyme. A new palace on the Charles River did not seem impossible.

Well, I thought, if Jamie makes his fortune this way, trying to save his brother's life, what is wrong with that? But I worried about the palace across the river. It was hard enough to think clearly about any of this without visions of gold rising in the dark. Matt and Jamie had incorporated their start-up a few days before, CNScience. They had other start-up names floating around, too, which they sometimes mentioned in passing: Uptake Pharmaceuticals, Neurologix. On my last call to Jamie to plan my trip to Newton, a new voice had answered the phone: Lizzie McEnany, another honorary Heywood, the daughter of one of Peggy's oldest friends. Lizzie sounded very professional and about sixteen years old.

"CNScience!" she said.

"What?"

"Doesn't that sound good?" She laughed.

Jamie owned the EAAT2 idea. He had made the first important description of the concept, and he would soon be filing for a patent on it. If anyone wanted to sell the idea to patients, they would need a series of patents, and one of them would now have to be Jamie's. Unfortunately, there was already one patent on the EAAT2 gene. The scientists who found the gene had patented it. That was a problem, Jamie said. "In a market that's worth only fifty million, you need to own the whole market or you're crushed. You'll never recover your costs. In pharma, it's all of the marbles or none of the marbles. And that makes it nasty. That's why drugs are so expensive." To make the package he was developing attractive to Big Pharma, or profitable to CNScience, Jamie had to try to put a big package together.

What if he went for the money? Or what if he did not, and then someone else did? He told me about his favorite Internet idea back at the Neurosciences Institute, his concept for getting commercials onto computer screens. He had walked away from that one when he went home to Newtonville. Now it was well funded. Someone else had gotten rich overnight with his idea. He had opened the *Wall Street Journal* one morning and there was his baby. "I'm pissed off about it!" he cried, and his kid's voice broke with indignation.

But Jamie was afraid that he would lose credibility with almost everyone—including me—if he did turn this into a race for profit, and he could not afford to lose credibility. The task of putting together something like this quickly was enormous. With his gene therapy, the paperwork alone would be monstrous, even though he now had

Paola's Canavan proposal to use as a template. Not long before, he told me, a company filing for approval from the FDA for a new drug had to send over four semitrailers full of data.

As we entered Newtonville, I asked Jamie how his parents were doing. He gave me a techie's answer. It was also a strikingly self-centered answer. "My mother is extremely sensitive to my state of mind," he said. "She seems to have an amplified pickup to my state of excitement or depression. My dad is totally pumped."

Jamie turned onto Mill Street and swung into the driveway of his parents' house. We cut across the backyard under a few tall trees. That September was warm in New England, and high above us I could see by the shadows against the sky that the trees' crowns were still full. Jamie picked up a sledgehammer from where it leaned against one tree and propped it against the side of the porch. The back door was open, and John and Peggy Heywood met us in their kitchen. Peggy was in her robe. They each welcomed me as warmly as if I were an old friend. Peggy told me that Stephen and Wendy were sleeping in the Cave. Peggy had prepared one of their guest rooms upstairs: Jamie's old room.

I could not fall asleep that night. Sitting up in the bed, I flipped the pages of a travel book that I found on a shelf, *The Happy Isles of Oceania,* by Paul Theroux, who had toured the islands of the South Pacific alone in a folding kayak. The book reminded me of my own adventures in the Galápagos—which I remembered now as simpler than this one. Once in a tent at a beach camp with no other human beings for fifty miles around, I had scribbled in my notebook, "To be awakened by the dawn and Darwin's finches."

When I told my mother about that adventure, back in her old kitchen in Providence, she shook her head. "What a romantic."

Sometimes now I wished I had never gotten into the subject of millennial medicine. I daydreamed about the Galápagos as if the islands were paradise. I wanted to go back to that tent.

When I let myself hope for Jamie and Stephen, I knew that I got carried away. It was a little strange for me to feel hopeful when almost no one else was. Even Matt During was more excited about his own brainchildren. Whenever Matt told him that the neurovaccine made the rats smarter, Jamie would protest: "You don't know much it scares me when you say that!" He did not want the treatment to do anything more to Stephen than what it had to do, he said: Save those dying nerves. If the neurovaccine changed the behavior of Matt's rats, who knew what it might do to human beings? "Memory is a *side effect!*" Jamie would yell, again with his kid's voice climbing and cracking on the high notes. "Memory is *toxic!*"

Money can be toxic, too. How distracting would it be for both Matt and Jamie if they were trying to weigh their options for saving Stephen and at the same time they stood to get rich quick? And what role did I play in this, their exclusive shadow and storyteller? Just by having signed on, by bringing the promise of the gold seal of *The New Yorker* to their project, I might help them save a life, and make their fortunes. Or, if things went wrong, I might help hurry a death.

After my first meeting with Matt During, I had stood in the living room in Providence, looking at my mother. She was sitting in the old wing chair holding the *Providence Journal* rigidly in front of her, which she did now for hours, staring at the same spot without moving either the paper or her eyes. I knew that if I let myself feel it, the excitement and the responsibility of the healer would be almost intolerable. And in the same instant, feeling that abortive rush, I knew from the very pressure of my excitement that it was wrong, that it was juvenile.

The real question was, what did my mother want to do? But that month she was suddenly too out of it to make the smallest decisions for herself. On that visit I tried to force the conversation by going over my parents' living wills with them. She had signed one a few years before—no extraordinary measures to keep her alive if her brain was so irreparably damaged that she could never again be herself. She seemed to be following the language as I read it aloud, but when I came to the line about brain autopsies she looked away. At the kitchen table, our eyes met, my father's and mine. *This is too much for her.*

Could I do it to her, when she had always been so wary and disliked science as far as I could tell, considered it something she could never understand, something for her husband and her sons to chatter about or her husband and his engineering friends with their perpetual talk of puzzles, problems and problem-solving, reasoning things to bits. Could I sacrifice her against her will? Just getting her out the door to the senior center a few blocks away from the house was a production now. During and his team would need to do at least one brain biopsy and possibly several before they prepared the neurovaccine. A brain biopsy is not a trivial procedure. And then the vaccine would almost certainly do her no good and might do her harm. Dad and I would be making her our little science experiment. It was repulsive to think about.

She was declining fast that month and my father and I were both upset. When I called home, oh, that repressive formality.

"How is she?"

"Well, she is not so well today, *Jonathan.*"

Right. Well, I'm glad I called, *Jerome.*

And god knows how my own voice sounded to him. We were not doing so well as a family. When I brought up the idea of a neurovaccine with my brother Eric in New York, I got a frown that I had come to call the Fortified Forehead. He and I could talk about almost anything but Providence.

I had felt that one moment of electric excitement, standing in the old living room by the wall of bookshelves that she had had built, and the old staircase, a little shabbier now, the whole room shabbier with the rugs taken up because she tripped on them, and the new railing running up the stairs. My mother sat bent over in the old wing chair, her wheelchair waiting next to it.

The rat with its whiskers twitching and its belly shaved and the golden spot, where the shot went in.

For what? Not for her. For the science she had never liked. It was too late to explain it to her. If she ever could have gone there, she could not now. She had been sick for a long time, and very angry and frightened. That night, in Jamie's old room, I remembered a scene that was, and I hope will remain, the worst moment in my life. Once, years after her fall in Vermont, but before I knew how bad she was—although my father knew—I was alone with her in the kitchen. She was chopping up a chicken at the cutting board on the counter with a sharp knife. I could see from her back that she was upset about something, and I went up quietly beside her and put my hand on her shoulder. She was hacking and stabbing at the chicken. I stood there at the kitchen counter with my arm frozen around her shoulder and watched her stab that raw meat with the knife. She turned to look at me. And I saw—taking it in slowly—that her face was livid, pale and flushed at the same time. I realized, incredibly slowly, that after all those years of too much painful sensitivity and insecurity and that weird secret hidden anger, she had now gone mad.

My poor mother was not Ponnie and she was not my mother. The magician of the kitchen had vanished, and the protective or overprotective mother and grandmother who held my boys' hands in New York City with what we used to call the Grandma Ponnie Grip.

"You fucker," she said. The voice was not hers. She growled a rage I had never heard before and words I never thought she knew. She cursed me and everyone in the house in a storm that was much larger than she was, some of it incoherent and some pretty clear. She was utterly lost from this world. If she had been in the room with me, she would have been more horrified than I was. But I kept my arm around her shoulder and she put down the knife.

Then all of us were standing together in the kitchen, my father, my wife, my sons, my mother, while she wailed a wail that meant the abandonment of the universe, the universe we build for ourselves and one another with such hard work for all our lives together, constructing and reconstructing the world of things as they are. It was a howl from a place where nothing is true, where nothing is the way things are. I see now what it was, what I felt then, though then I could only feel it: the horror of a world without supports—for me, for her, or for any of us. The horror of the unbuilding of everything.

Lying in bed, wide awake at two or three in the morning, I read *The Happy Isles of Oceania* by Paul Theroux, whose journeys in the folding kayak had not gone very happily.

Jamie Heywood's adventure was an outsider's race, I thought, but it was in the style of the whole of biomedicine in the last year of the millennium. It was part of that sense that life could now be engineered and health be engineered; that the study of life and the treatment of disease were passing into problems of engineering. "Given such a view," Oliver Sacks writes in his book *Awakenings,* "one can conceive the possibility of affecting a single point or particle, without the least effect on those surrounding it: One would, for example, be able to *knock out* one point with absolute accuracy and specificity."

Illness makes us all yearn for that kind of magic cure; even those of us who are well feel the pull of it, because all of us share the feeling that we have fallen somehow from a state of perfect health and wholeness, a state of grace we once possessed. Sacks writes, "We had something of infinite beauty and preciousness—and we lost it; we spend our lives searching for what we have lost; and one day, perhaps, we will suddenly find it. And this will be the miracle; the millennium!"

Jamie had chosen an injection of DNA as the Elixir of Life that could save his brother and his family. It was at once a promising idea and part of the general genomania. The worldview of the genetic engineer lends itself to a kind of mysticism about the world, a vision of reality as a network of bits and pieces. "The therapeutic correlate of such a mysticism," Sacks says, "is the notion of a *perfect Specific,* which has exactly the effect one wants, and no possibility of any other effects." Sacks gives as one example the drug that Paul Ehrlich tried in 1909 for the treatment of syphilis, a drug that its promoters and enthusiasts called "The Magic Bullet." A Magic Bullet is the ultimate techie dream and Jamie was building it. Magic Bullets are sacramental medicines, Sacks writes, in which we can see "the amalgamation of genuine needs with mystical means, the mistaking of an infinite, metaphorical symbol for a finite, ingestible drug."

What if I fired that bullet at my mother? My God! Science with a vengeance!

Sacks writes, "We may expect to find such ideas most intense in those who are enduring extremities of suffering, sickness, and anguish, in those who are consumed by the sense of what they have lost, or wasted, and by the urgency of recouping before it is too late." A man in such a state becomes desperate for regeneration. "And it is at this point, when he is searching, here and there, with so painful an urgency, that he may be led into a sudden, grotesque mistake. . . ." Then he and his doctor may plunge into unreality together.

Yes. It was the sensation, the conviction, of a grotesque mistake that kept me awake that year in the late middle of the night—as if hope itself had become a nightmare from which I was trying to escape.

We were all in that state and so was Jamie, who was besides a wild-eyed entrepreneur in a wild-eyed year. Were we plunging into unreality? There was no way to find out except to try the experiment. Turning the light on and off in Jamie's old room, trying to close my eyes and keep them closed, I worried about Stephen, and my mother, and I wanted Jamie to fire his Magic Bullet—as if one superhero with one bolt from the future could save us all.

{ T W E N T Y - S I X }

Denial Is in Egypt

By the time I made my way downstairs in the morning, John had already left for MIT on his bicycle. Peggy served me a wonderful breakfast in the sun porch, just off the kitchen, the room they called the solarium, and she and I fell into a long, comforting talk. By my second or third mug of coffee I felt much better. I told her how impressed I was that Jamie, an outsider to biology, could make a plan to save Stephen and figure out how to make it work.

"That's what engineers do," she said warmly. "You don't need A's in physics to be an engineer. You need to know just enough about how it all works to use it to solve your problem."

While we talked, Stephen and Wendy stole up from the basement and tiptoed through the kitchen, trying not to interrupt. Stephen had the slacker attitude in his very posture, a slight sort of adolescent slouch that went with his T-shirt and stubble, and had nothing to do with ALS. He was still a well-built man, though both his arms looked as if they were losing strength. Wendy was barely half his height. She was blond, round-faced, and pretty. She was on her way to her job in Cambridge at a Harvard biological laboratory.

An hour later, Stephen and I made plans for the conversation I wanted to have. For the *New Yorker* article, I wanted a big deposition: his telling of the story. We were talking down in the Cave. A big canvas of Stephen's leaned against a wall: a Swiss Army knife, in oils. Stephen had painted it at home one summer when he came back from Colgate. "I don't care for emotional or expressive painting," he said. He showed me a small watercolor he had made in college, a close-up of a pulley.

Neither of us was feeling comfortable about our day together and where our meeting was going. Neither the carpenter nor the science writer was prepared to work on a story this emotional. Looking at that pulley, I could see that between the two of us, we were going to have to figure out a way to get at things from a distance, from a tangent. I could not just ask Stephen, "How does it feel to know you are dying?"

But I liked Stephen very much. He was not feeling sorry for himself. He was not even particularly preoccupied with himself, even then. Stephen had his interests and they were not Stephen. He asked me about my home in Bucks County. He knew the place because he had worked on the Victorian in Yardley for Jamie and Melinda, remodeling their bedroom. I told Stephen that too much of the county's farmland was turning into acres of schlock McMansions. Stephen said, "To me, that's the great American tragedy, the loss of open space." And he added, "I'm sure other people would find other things to feel sad about."

That day Stephen had to go to the hardware store and a lumberyard in Wellesley. He was making a bookshelf for his parents. He had to buy a saw, and some veneer, and one or two other things. We decided that he would tell me his story while he ran his errands, and we got into the truck.

"OK, so start from the beginning?" he asked. "Just battle on?" And he chose to begin with the story of his first house, the project in Palo Alto. He was just digging the foundation when we stopped in the parking lot of a GNC. Stephen was taking Metamucil, Creatine, and riluzole, still the only drug the FDA had approved for ALS. Riluzole is a glutamate blocker. If Jamie's gene therapy project worked, it would install billions of EAAT2 pumps in Stephen's synapses, and they would pump away that excess glutamate. Stephen also picked up some Coenzyme Q-10 dietary supplements. Above the shelves, there were big blowup black-and-white photos of smiling, healthy, happy people.

"Incredibly expensive," Stephen said as he limped to the counter. "Insurance doesn't pay for a cent of it. But everyone with ALS takes them. You know, what the hell."

At the counter, Stephen dropped his credit card, keys, and wallet on the floor. He had always been klutzy, he said, but he was getting worse. "What I really should do is mail-order my stuff. I just haven't gotten that organized. I keep saying that I need to get myself a daily regimen that involves the vitamins I take every day and all the medicines, as well as a little bit of time for meditation or exercise. But I have not managed to do that yet."

"Well, everything's in flux right now."

"Yeah. But I'm not so foolish as to assume that therefore it will not remain so!" he said, with a dark laugh. "I keep telling Wendy that things will become quieter eventually, but then I realize that may not actually be the case. So perhaps we should just learn to steel ourselves to the inevitability of chaos."

Back in the truck, Stephen returned with obvious relief to the challenges he had faced in the foundations of his first house. "Very, very complicated," he summed up cheerfully. "So that was a lot of fun figuring that out."

He drove around Newton, Newtonville, Newton Center, Newton Corner, and West Newton, stopping here and there to show me his favorite Victorians. By the time we reached the hardware store, Stephen was explaining the landscaping of his house. From the way he relived it day by day, I could see that it really had been a work of art more than a house for him. He was not an ordinary carpenter. More than a little of the academic had rubbed off on him even as he rebelled against it. Newton's apple had not rolled that far from the tree.

Before we got out of his pickup truck, he confided to me that he still felt very insecure when he went into a hardware store or a lumberyard. "They're old and gruff and they *know,*" he said. Maybe it was an age thing, or maybe it was a blue-collar thing. This surprised me, because he talked blue-collar himself. But in the store I could see that he really was from a different culture—my culture. He was comically polite and professorial in asking his questions, even apologetic. "Sorry. Excuse me. One more thing—oak laminate?" he asked a spindly teenage clerk. Then he apologized to the kid again and asked if he needed something special to trim the laminate. "Nah," the clerk said brusquely. "Just use a razor knife." Stephen bought an edge trimmer and a roll of cherry veneer for the bookcase.

Now that his errands were done, we drove around some more in the truck. I thought it was getting time to talk about his illness, but he turned instead to the brass hardware that he had bought for the windows of his house in Palo Alto.

Finally I said, "Maybe that's enough about the house."

"OK," Stephen said. He gave me a sheepish and awful grin. At last he began to tell me about the weakness of his hand as he tried to turn the key in the front door, and the doctors' tests, and all the rest of the saga. Of all the topics in the world, this was the one he liked least.

"It's weird, I don't know," he said. "You always hear horrible stories about the doctors saying, 'You've got two years to live.' But I never felt like I was thunderstruck or anything like that, and it's not like I expected anything like this to happen to me. I guess I was like 'Oh, OK. Things are gonna change a little bit,' but it wasn't like 'Holy shit, my life is over.' I never had that feeling that a lot of people report, that *bang,* you know? I don't know why that is. I guess with this disease, it's a little bit like carte blanche—you can do anything you want. I mean, it sounds terrible, but you're like, 'Oh, my whole life has changed. What do I do now?' I was feeling healthy as a horse and I was like, 'Well, shit, I get to screw around for like five years.' You know?

"My real first impulse is, 'Oh, I'm gonna screw around for the rest of my life.' So I'm not really involved in what the foundation is doing. The fact that Jamie's doing this—it's like he's taken my responsibility. I mean, I really do feel like that. And he's not only doing it well, just doing an incredible job. So I feel liberated. If when all this is said and done, you know, this is how people remember me, they can say, 'Way to go, Steve, look what happened.' Whereas I won't have had that much to do with what happened."

"You won't have had much to do with it?"

"I mean, when I'm dead and gone, and people are looking at what I've done with my life, I can add this on, even though *I* didn't do it, exactly. The amount of energy that Jamie is putting into this constantly—and Melinda, too—is in itself phenomenal," he said. "Because they don't have to. I mean," he said, with a painful humility, "I know that's why you're writing the story."

He parked the truck outside another house he admired in West Newton. The street reminded me of College Hill in Providence, with Victorian masterpieces on every block. Sitting in the truck, staring straight ahead through the windshield, Stephen said, "Jamie's not always that focused. I mean, this level of focus is incredible. So I can think a little bit that he's doing it because it's exciting and he loves it. But that's not all, that's not it.

"So you think to yourself, what *is* his driving force? You can tell by his time line. He's not interested in doing stuff that takes a long time, and there's a reason for that. I mean, he's trying to do it *bang!* And obviously he's not making any money or whatever. Although it's not gonna put him into the poorhouse. But, every day! If *I* had to get up and do something every day, I'd have to have a pretty good force driving me."

Stephen glanced at me. I saw that he was trying to make a speech for the record, to do what he had to do for his brother, to say, a little stiltedly, the one thing he least wanted to say out loud.

"And so, that's moving to me," Stephen said. He seemed to be forcing himself to go on, even though he hated schmaltz. "It's just like asking someone to hold up a rock for ten days. And you say, 'Well, sure, I could do that for ten *minutes*—anybody could do it for ten minutes. But for ten days, it's impossible!'"

"And yet there he is."

"Yeah. And if you said, 'Hold it up for ten days, or your brother *dies—!'* "

With that, Stephen's face suddenly contorted. "I mean, you know, he's *doing* it, though." His mouth worked. He choked up. "*Oy vey!*" he said.

We sat in the cab of the truck. Stephen shot a sidelong glance at me.

"I shouldn't say *oy vey,*" he said.

"Why not?"

"I'm not allowed to, necessarily. I'm not even sure what it means."

He was not Jewish, but he had gotten the cry in the words just right. I said, "It means, *Oy vey!*"

It was up to me to follow that with those reporters' scene-of-the-disaster questions: *How does it feel to know you are dying?* It was my job that day to ask them, but I just could not do it. I think we were both happy to go back to talking about houses. We felt even more at ease when we got back to Jamie's Victorian, a few minutes' drive from their parents' house on Mill Street. Jamie joined us, and he and Stephen showed me the Victorian and the carriage house. The carriage house was set back from the main house about fifty feet in a patch of woods. It looked to me like nothing more than a half-ruined garage. There were outcrops of stone around the carriage house, gray conglomerate boulders with big cobbles, like storm clouds full of hailstones. Jamie pointed them out to me as proudly as a realtor. Roxbury puddingstone, the State Rock of Massachusetts. A stone with a place in American history. Children around Boston have been climbing on Roxbury puddingstone since before the Revolution. Local churches are built of it. A thirty-ton monument of Roxbury puddingstone was sent down by train to the battlefield in Gettysburg to commemorate the 20th Massachusetts Regiment.

"I'll let Stephen tell you," Jamie said, and went right on talking as we circled the carriage house. They would work with the Roxbury puddingstone in redesigning the house. "I love this idea that constraints make architecture interesting."

"Jamie's stealing all my quotes," said Stephen.

"These are all Stephen's quotes."

"We just want to achieve a separate feel," Stephen added mildly, nibbling on a twig.

"I love how this building sits inside this pocket of rock!" Jamie said.

"And it'll do that more when we're done, actually, Jamie."

Stephen and Jamie trudged around the carriage house talking possibilities. They saw dormers on three sides. They would raise the ceilings and put in skylights. Stephen said he wanted to make the first two floors wheelchair-accessible. But he wanted to design the ramps in a way that would be aesthetically pleasing, he said, "noninvasive."

Jamie laughed. "And all the while we have to figure out how not to lose money," he told me. "ALS has economic consequences. The expense starts at $10,000 a year. Ends up $170,000 a year, with intubation and round-the-clock care." Intubation means a plastic breathing tube in the windpipe, and an artificial respirator.

"Assuming you want that," Stephen said. He leaned against a tree behind the ruin, near a big storm cloud of Roxbury puddingstone. "My feeling is, you get to that point and you don't want to keep going. Shit-piles of family around by then." Absentmindedly, as he spoke, he removed a bug from Jamie's collar. Jamie endured this unconscious brotherly gesture and endured the drift of Stephen's thoughts with a kind of tender blank-faced look. Suddenly we were having one of the conversations that Stephen and I had avoided in the truck. My father and I were talking about this now, too—to intubate or not to intubate. I thought my mother's living will made her wishes clear. My father did not. Now Stephen was telling Jamie what he wanted.

Stephen slouched against the tree, looking off toward a few crows that hopped on the grass at the edge of the woods.

"How do you know that now?" Jamie said. "Intubation could cost you a night's worth of medical intrusive care, and buy you a year and a half of high-quality life."

We listened to the faint suss of traffic and the breeze in the trees. A dog barked. Jamie broke the mood and turned back into a Realtor.

"This is a great dog neighborhood!"

The carriage house had been abandoned a long time. There was an old rotten canoe propped on its side against it. The door was white, weathered, peeling paint, with gray wood showing through. "We'll save the big front door," said Stephen. "And maybe put it on a motorized slide. Computerized."

"Computerized?" Jamie said.

"Computerized. Oh, sure. I'm going to computerize the house. And I'll just shout out, 'Door,' and it'll slide right open."

Remembering the scene now, I realize that the Heywoods could not have talked much about the future. Stephen at least had thought very little about it. If he had, he would have remembered that he was going to lose his voice.

Today Jamie often talks to ALS patients who are near the start of their road, where Stephen was then, people who are still busy just enjoying life—and why not? "Denial is a fine place," Jamie likes to say. "It's in Egypt."

Inside, the carriage house was a wreck. A lawn mower was parked just inside the doorway. Beyond that, the floors and walls were dark and in ruin, with cobwebs draped and drooping everywhere. Jamie and I went up the steps first and waited on the second floor while Stephen followed very slowly after us. He kept half-stumbling and apologizing to us for the scares. His right leg was trembling with what neurologists call fasciculations. Rock climbers call it "sewing-machine leg." Stephen's fasciculations made the steps an adventure for all three of us. They were steep, and the old treads had been worn narrower from overuse.

"The stairs will have to be a little more code next time," Stephen said, and looked around. "This is rat shit, Jamie."

"There's tons of rat shit," Jamie agreed.

"That's really unpleasant," said Stephen happily. "Look at the paneling in that wall. That could make some great paneling on the outside wall. Then this could be the living room, kitchen, dining room."

"But all open."

"But all open. Upstairs, bedrooms. And downstairs, a playroom for kids. This could be a door right here." He gestured at a bare window darkly shaded with cobwebs.

I saw nothing but ruins and junk.

"Nice roof," I offered.

"Nice roof," Jamie agreed, emphatically. "Yeah, this is a really nice project."

Up a ladder, in the middle of the hayloft, a desk that once belonged to a lawyer was still covered with his papers. We picked up correspondence from 1938. Someone must have decided to close up the place without touching the desk. The papers themselves were covered over with dust and cobwebs. I stooped and picked up a book from the floor. It was an ancient Funk and Wagnalls *New Standard Dictionary.* Much later on, when I was looking at the carriage house with Melinda, I said, "Some old lawyer left his last case open."

"He wasn't an *old* lawyer," she said. "He died young." His widow had stayed on in the house for sixty years after he died. She lived alone all that time.

But Jamie did not mention that. He was selling me the house, and his plan, and his brother, and their amazing story, all at once. The floorboards were rotten and they shook as we walked. "Look!" Jamie said. "This is the coolest music stand on earth. Cast iron. Weighs about eight tons. Maybe it's an artist's easel. This was all really nice furniture before it was destroyed. This will be a great space. There's the original hayloft door. Stephen will make it glass—you could open it up for a breeze in the room. I always wanted to do one of those Pennsylvania barns. Stephen is an amazing craftsman. He doesn't like architects. He thinks they draw something and walk away."

"My quotes."

Slowly we made our way back down the steep and narrow stairs.

"Needs a very large Dumpster," Stephen said, looking up at the place from outside. "But the fun part will be pulling out the wood and salvaging everything." He pointed out details he liked, as he had been doing throughout the villages of Newton all afternoon. "Wonderful diamonds over the windows. Crosses. All real-dimension lumber, not like the crap they make today."

We strolled back to the main house. "This place is major TLC, too," Stephen said with satisfaction, scanning the eaves. "Long-term. It took me how many years to do my first house?

"Hooh! I'm tired," he said.

The carpenter and the engineer leaned back against Stephen's truck with their arms identically crossed on their chests, wearing identical grins, looking up at the Victorian.

{ TWENTY-SEVEN }

The Jamie Factor

At the foundation one afternoon, Jamie and Melinda and Jamie's staff of two met with a consultant he had hired, Katherine Evans. She was his new project manager and she also saw herself as a tough-love mentor, he told me. Jamie had warned me that the meeting would be a review of everything they did not know. "So it's going to be a brutal and depressing meeting."

It was a Friday afternoon, two o'clock, a time when most offices are slouching toward the weekend, when the meeting began. Jamie and Melinda were setting the table for lunch in the dining room of the Victorian, within sight of their desks and computers. The dining room was unfinished but Melinda had added quirky touches to decorate it. Black cats cut out of tin perched or prowled or draped themselves on each white lintel. Sun shone through wide-parted, many-splendored drapes in the bay window, an Indian print with a gold border. Against one wall, a table made of an old Singer sewing machine held green plants growing in baskets. Displayed on the Singer was a lighthearted coffee-table picture book, *Master Breasts,* paintings of breasts by the Old Masters. Under the chandelier, Melinda had spread the table with an old hippie floral tablecloth that she had found in Provence. The whole house was Victorian with touches of her child-of-gypsies style.

Katherine Evans was middle-aged, Chinese American. She had

been project manager in the development of a genetically engineered version of erythropoietin, a human protein that boosts red blood cell production (it is used for patients on kidney dialysis). Katherine had worked on erythropoietin at a biotech called Genetics Institute, racing against a rival biotech, Amgen. There had been a highly publicized patent battle. Amgen had won, and erythropoietin was now famous in the overlapping worlds of medicine, biotech, venture capital, and Wall Street as biotech's biggest success story: Worldwide it was the best-selling product in biotech history. That year, according a press release on Amgen's Web site, the company expected its sales of erythropoietin to reach almost half a billion dollars.

Katherine was now coming off another drug development failure. She was consulting while she looked for her next job. The new failure was still raw for her. As a consultant, she had a faint aura of grief and bereavement, of enforced singleness.

An old friend of hers in the industry had called and asked if she would give Jamie some advice. "When I first got the phone call, I thought—what has Bruce gotten me into?" she told me softly, and I nodded. I felt the same way about my friend Ralph.

At that afternoon's meeting, Katherine was supposed to give Jamie an overview of the drug development process. She would help him put together a master plan—how to get approval from the FDA for human clinical trials. Before we started, I asked her to give me a quick introduction to drug development.

"You know Stephen is in the next room," she warned me quietly and urgently. He was playing a computer game called Starcraft. He had no energy that afternoon, he complained to us through the open door. The game was not going well.

Katherine explained to me how different Jamie's project was from standard drug development. Usually, the story of a new drug starts with a drug company. It could be a biotech or Big Pharma. The company makes the product. Once they have their product, they need assays to define what the product is. They need to test its purity and stability. Then the substance is usually tested in vitro for activity. And then in vivo in small animals, a few mice or rats, for toxicity. Not hundreds of animals yet. Then in more mice, and rats. Then in dogs, rabbits, pigs, monkeys. "Don't write dogs!" she cried. With animal tests, they look at safety and efficacy. They test different formulas of a drug, and when and how it is delivered. They test that the agent is safe and the route of administration is safe.

Next, there is a Phase 1 study with volunteers. Usually the test involves a dozen people or so, to see if the drug is well tolerated. In Phase 2, the drug testers find out about dose and regimen. They find out what dose works best. They find the half-life of the product. They do a number of Phase 2s, with different patients. They may do four Phase 2s. This can take years.

Then they design Phase 3. Here they want to hit a home run. The Phase 3 trials are double-blind, placebo-controlled. "You design absolutely the best, tightest study you can. So that at the end of the day you can apply for product license approval from the FDA. Or new drug approval."

Normally, Katherine explained, all this would be done within a company by specialists who were used to working together as a team. But of course no one else was working on Jamie's idea—or, as Katherine put it in the jargon of biomedicine, "on this particular indication." And normally the total cost of the project would be tens of millions of dollars. To bring out a new drug, a giant pharmaceutical house may spend almost a billion dollars and ten years.

There it was again: the voice of the way things are. Whenever I talked with people besides Jamie, I had to try to sober up, and I wanted to less and less. When I talked that month with Jeff Rothstein in Baltimore, I told him that I hoped to end my *New Yorker* profile of the Heywood brothers with the excitement of the trial.

"In reality—" Rothstein began, and backed up. "I don't know how much you know about trials. But the first time you stick something new into a human being, the only thing you really care about is, is it safe. I've had to admonish Jamie greatly about this. 'You have hopes that are different from what we have as responsible physicians.'

That is, the first time you try something new, you gotta know, can I do this without harming someone? And that's all this first trial is. The reality is, we're trying something that hasn't been done. I mean, in the very crudest sense, we're going to stick a needle in your spinal cord, inject something, and hope that it helps you. The first thing we want to do is make sure that it doesn't hurt you. And that's what a Phase 1 trial's about. Can you do whatever you want to do safely? And if you can do it safely, the next step is to say, 'Gee, what *happens* when you do it?'

"There's a long series of hurdles to make a drug. This is the first one. Can we put in the DNA and not hurt someone, sticking a needle in the spinal cord?" Rothstein was afraid that the operation itself would be hard for a surgeon to bring off. It would be much trickier than sticking a needle into the brain to cut out a tumor. "In your brain, you have a lot of leeway. But your spinal cord is the thickness of the tip of your index finger. Not big. Not big. And a mistake there is catastrophic. Especially at the level of the cervical spinal cord. That's sort of the equivalent of Christopher Reeve's injury. The last thing I want to do is have anyone undergo this procedure and come out paralyzed from the neck down."

Katherine and I both knew what an extraordinary maze she had just diagrammed on her yellow pad with her mechanical pencil; and we were both old enough to feel parental and protective as we watched Jamie and Melinda lay out cartons of take out Thai food. They made a fine sight, Jamie with his thousand-watt energy, his Boy from Tomorrow look, Melinda rubbing her pregnant stomach and nibbling on a rice cake. She was not showing yet.

"You're smiling," I told Katherine.

"I smile a lot," Katherine said.

Jamie cut in. "At the Heywood table the etiquette is, fight. Fight for what you want."

Everyone at the table dug in but Melinda, who had no appetite—

her morning sickness lasted most of the day. While we ate, Stephen opened the door of his computer room and passed through the dining room. “I won,” he said gloomily. He was not hungry, either. He went out to get his pickup truck and buy some more wood. He was going to walk a mile or two to his truck, which was back at the house on Mill Street. He refused Robert and Jamie’s offers of a ride. After he had gone out the door, a silence weighted the air. What was in that silence? I wondered. It contained everything that everyone in the room was choosing not to say.

After lunch, Jamie began the afternoon meeting. Robert Bonazoli stood at a flip chart with a list of duties, which he had spelled out all the way down to who ordered the bottled water.

First he talked about their temperamental printer. It gave them error messages, it talked back: “Too hot or humid to print.” They had to buy an air conditioner for the printer. “I love our printer,” Jamie said. “ ‘I don’t feel like it right now.’ ” Robert said Melinda was working on taking care of the printer. “And the pregnancy thing she’s working on full-time,” Robert added.

Robert had spent a lot of time on the next fund raiser. In December they were planning a joint fund raiser with the Massachusetts chapter of the ALS Association, ALSA: a Christmas concert at Jordan Hall of the New England Conservatory. The orchestra, the Longwood Symphony Orchestra, was made up entirely of doctors, nurses, and other medical people. They would play Beethoven’s Fifth. Jamie and his foundation had now built their mailing list up to 9,700. They were calling it “A Concert in the Key of Hope.”

Next, Lizzie reported that she had found some great sites on the Web, including a new journal of gene therapy. She had subscribed. Now they could just double-click on results and get them. There was a team in Switzerland just then that was injecting genes into the human spine.

I glanced over at Stephen’s room. The door stood open. I was glad he was out.

Katherine urged Jamie to make an informal call to the FDA and talk over his plans. But Jamie did not want to do it. "My bullshit capability's not quite up to calling the FDA and saying, 'I want to do this,' " Jamie said.

Next they turned to the toxicity study. They were testing each version of their viral capsule on mice and rats. A woman from Toxikon, a huge animal lab that does a lot of studies, had volunteered to consult on that.

"Awful name," Jamie said. "Almost as bad as StressGen."

Jamie shepherded them onward through their list of to-dos. "Develop action plan," he read aloud. That was on Jamie's list. "I *thought* about it," he reported. General laughter. "Actually, we got nearer than that."

Jamie said he was working on their medical advisory board and he was trying to land one more celebrated biologist. "The hook is set," Jamie said. "I want to let the fish swim around a little longer before we yank 'm." He asked Lizzie to try to set up their next meeting. "I'd much rather do the sale in person."

Although he was focused on the gene therapy, he was also eager to back the ALS vaccine. "We gotta get on the ball on this," he said. "As soon as the vaccine comes out, I want it splashed across the world that we—I want to *own* it."

Katherine asked where Matt During's neurovaccine paper was. Jamie explained that Matt was getting critical reviews. "He's got to do more experiments."

"What are they objecting to?"

"It's completely new. His last reviewers said, 'This contradicts everything we think we know about the immune system.' "

I saw Katherine's face register that.

Jamie went through the long and still-growing list of treatments that he and Matt had invented so far. The two cowboys' capacity for monumental dreams had made a nightmare of their to-do list. The gene therapy alone was getting very complicated. "We've got Uptake," Jamie began, meaning the glutamate uptake project, his EAAT2 gene therapy. "Two different possible technologies, three different promoters, and a couple of switches. So twelve different molecules we can use. Matt and I are talking about how we can narrow them down."

"Good," said Katherine.

"The safest molecule we can't make," Jamie said. "It's too big."

Stephen chose just that moment to walk back into the house. He stalked through the dining room silently with a glower and closed his door.

After a pause Jamie continued. The talk about the construct got technical. They hoped Dave Poulsen at Matt During's lab was on schedule for October delivery. "Thing is, the technology keeps breaking," Jamie complained. With gene therapy, most of the trick is getting a useful gene to insert itself in the right place and turn on. But gene therapists also want to be able to switch it off, if they have to. Genetic switches are finicky.

"We need to know the neural consequences," Jamie went on. "We're trying to decide, do we do the brain and spine or just spine? And we need to decide which of these twelve molecules we're going to use. We could run a monkey trial—two monkeys with spine, two monkeys with brain."

Lizzie sucked on a lollipop and scribbled minutes.

Rob tried to sum up the state of the construct. "So the point is, for me, we have many versions of the gunk. How do we boil it down to decide which gunk to use?"

He meant they had too many options, too many ways to put the construct together: too many different kinds of switches, enhancers, promoters. It was as if they had too many pieces in their LEGO set.

"Well, what's really annoying is, I would rather inject the switched form into Stephen," Jamie said again, "because then we could switch it off!"

"Maybe we have too much to choose from," Katherine said.

I knew all this technology from the fly biologists I had hung around with for my last book. The fly guys had also spent their time designing experiments in which they inserted genes into their subjects and switched them on and off. I scrawled in my notebook, "What a difference a day makes." In his fly lab at Cold Spring Harbor, Tim Tully had once said to me, "Thank God I just work on flies." In another fly lab I had seen an old cartoon on the wall: a group of masked and gowned surgeons and nurses around the operating table, preparing for the first incision in a patient with six legs. The caption said, "He's a very sick fly."

I wondered what this meeting was like for Stephen to overhear or tune out. In his place I might have played computer games, too.

Jamie taped up a new master plan on the wall: "UPTAKE: GETTING TO THE GUNK!" The gunk was the stuff they would inject into Stephen's spine. Jamie was so excited to be approaching his goal that he seemed to forget that gene therapy had never worked. "This is totally cool!" he cried.

"Now, 'Uptake—Manufacturing,' " Robert announced.

"Could you not think in shorthand?" Jamie asked.

"I can. It's just not your shorthand. I'm not in your head, you know."

"You guys have been together too long," said Katherine.

"We have. I think we need marriage counseling," said Robert.

Now they talked about biotechs in Germany, scientists in Switzerland, regulations and their rationale, transporting the gunk across state lines. Katherine Evans kept coming back to the rules of the game. "It's not something that's tweeze-apartable by logic," Katherine explained, when Jamie protested. "It's not about common sense. It's what the regulations are."

They would try out the procedure on monkeys first. They would get the monkeys from medical supply houses. Unlike the rats and mice, the monkeys would not have ALS. But monkeys have much more humanlike spines, brains, and immune systems. "I'll tell you, I used to have moderately mixed feelings about animal tests," Jamie said. "But when you're trying to decide what to do for your brother, your perspective changes quickly." He thought the injections would probably go into the spine, not the brain. "You don't die if you lose your spine."

"What does it do in the brain?" asked Katherine.

"You can lose a lot. So I'd rather do the spine. If you dump into the brain, you've changed a major neurotransmitter." He meant glutamate. "I don't feel good about that. You've changed how you think. I guess we're going to go spinal."

"Isn't this about as much as we can do today?" Katherine asked.

Jamie looked at his watch. "It's still twenty minutes to five," he said, in a tone of mock reproach.

"Each one of these is a lot of work," Katherine said, pointing at the long list of to-do items on the chart. "When you get down to the nitty-gritty and the nitty-gritty beyond that, it's a lot of work. Which translates to time." She asked Jamie how much time he thought it would all take.

"I think this is two people in Matt's lab for eight weeks, based on Matt's infrastructure."

"If this was a real clinical trial for Merck," Katherine said, "it would be thirty people or so." Although Jamie talked in wilder moments as if he had a hundred, in reality he was relying on the people in that room, the reluctant postdoc Dave Poulsen at Matt During's lab, and a few other postdocs in Baltimore and Auckland.

Jamie told her he thought they might know what the gunk should be by November 1, a little over a month.

"You know, there is a thing about haste, in drug development," Katherine said. "There is economy here," she said, pointing at the neat flowchart she had drawn on her yellow legal pad, when she laid out the standard path a drug takes from invention to delivery. "Economy in doing one step at a time. As opposed to doing everything in parallel." She had a solid, level, truth-talking tone. She was asking sharp questions, again and again. A middle-aged product manager has met many cowboys and self-made matadors, and knows how most of them end up.

"Put down November 15, we start safety," Jamie said. "That's pretty aggressive," he acknowledged to Katherine.

"Yeah," Katherine said, as if he had just stated something too obvious to mention.

"I can do it."

"No problem for *you,* Jamie."

They both looked at the list they had just made. I could see what Katherine meant. The list looked impossible.

It was 4:55. "We need to do assignments now," Katherine told Jamie.

"In the last five minutes," Jamie noted.

"Jamie's going to do it all!" Katherine announced. Everyone laughed and the meeting broke up. While Robert gathered up the loose papers, Katherine and I talked quietly at the far end of the dining-room table. "They're not full of self-pity," Katherine said. "If they had that element of self-pity, you could almost not stand it. I like them. I feel maternal. I like their spirit. I think Jamie will always find people to help him.

"But it's a lot of work, it's a lot of politics, a lot of sensitivities, and we don't know what the results will be. I think Jamie is in a unique situation. Because he's not a scientist, not an ALS researcher, you could quibble whether he's picked the right project. Well, I think that's almost beside the point, if you know what I mean. Why would a layman be able to pick something to do that the researchers in the field didn't pick?"

Melinda overheard that. She was walking through the dining room with her wet hair done up in a towel, rubbing her belly and nibbling another rice cake. Melinda was belly dancing at Karoun that night, in spite of her sick stomach.

"The answer is that it's Jamie!" she said. "That's always the answer! Really!" She stretched tall, shooting both arms up in a *V* in the open doorway, and shouted, in mock-cheerleader style, "It's the Jamie Factor!"

Katherine Evans disappeared from the scene soon after that. She was out of the picture—or at least, she was out of the picture that Jamie was trying to paint for me. I never heard him mention her name again.

{ TWENTY-EIGHT }

Generation X

On Saturday night, Peggy served a minor feast in the solarium. After dinner, Jamie and Stephen put their long legs up on the ledge of the wainscoting and faced each other like bookends from opposite ends of the table. They talked with John and Peggy about tools, houses, real estate, and the stock market. Stephen found it hard to imagine that the market could ever go down.

"It can happen, children," said Peggy.

"I don't doubt that the economy can do other than expand," Stephen said, "but until it happens I won't believe it." He had just read an editorial in the *Boston Globe* that made him happy. It was a defense of Generation X. People laugh at us, Stephen said, people make jokes. "And yet we're the ones driving the economy!"

"Oh, yeah," said Jamie, sarcastically.

"Well, it's true!" Stephen said, his voice climbing an octave just the way Jamie's did when he got excited or indignant. By then the giddiness of the stock market rise was beginning to seem like the new state of the universe. Here and there, money people were talking about a bubble, but software engineers in the Valley were holding on to their stocks. "Nobody wants cash: it's too final," one Silicon Valley investment analyst had written that summer in the *Wall Street Journal.* Why trade stocks for dollars? Behind the dollars stood Uncle Sam—just old Uncle

Sam. "Internet paper, on the other hand, is backed by smart entrepreneurs who work like dogs through the night to change the world." The engineers were bringing us the future in the fall of '99.

During Stephen's speech, Jamie stared at him from the other end of the table. Stephen caught his look and flushed. Between the two of them, it was Jamie who represented the smart entrepreneurs who worked like dogs through the night. Stephen was on the *whatever* side of Generation X. He was the grasshopper among the ants. He had built his first house with credit card debt, two mortgages, loans from his brothers and his parents. He confessed that he had even bought his Harley when he was in debt. Then one month before he finished his house, the Harley was stolen.

This was Peggy's cue to cry, ritually and ruefully, "It was a miracle!"

I told them how much I enjoyed these meals with them in their wonderful solarium. Stephen said he once had plans to renovate an old farmhouse in Vermont. He made blueprints for a silo, a greenhouse, and even a solarium like the one we were sitting in. With innovations in heating and cooling, a glassed-in porch could work even in Vermont. His solarium design would stay warm in the worst New England winters. "This is not that crazy," Stephen said.

"This is not crazy?" said Jamie.

"It was brilliant," said his mother.

"Shit, when's your birthday?" Stephen asked Jamie suddenly, looking stricken. "November 4? *October* 4?" His voice climbed an octave again. "Shit! What do you want?"

"A router-planer." Jamie went back on the attack. "What the hell are you doing renting a fourth-floor walk-up?"

"He's an optimist, and so am I," said Peggy, warmly. "Besides, he needs the exercise. Don't you, Stephen?"

The four of them talked late while I listened, admiring the gallant way they all held up under the weight. There was the weight of hope and there was the weight of knowledge—the weight of the way things are. The odds against Jamie's plan were so long that hope hurt: Hope almost felt like the heavier weight. When the dinner-table talk finally broke up that night, Stephen apologized to me for the intensity of the story that they had lured me into.

"I'm just passing through," I said. "If you can handle it, I can handle it."

"We've adapted," Stephen said. "You're going to come out of this with the bends."

It was Saturday night, and Jamie and Stephen decided to make it another Boys' Night Out. They drove back to the Victorian that John and Peggy had bought them. Every Boys' Night Out they played Quake on the foundation's computers with Robert Bonazoli.

Since I was not interested in computer games, I stayed at John and Peggy's. We cleared the dishes. Peggy wondered why it was almost always boys who played those games. It was very mysterious. Her sons had been playing them forever. Stephen had taught Dungeons and Dragons at an arts camp in Newton. Jamie and Ben had run the group. That was in pre-computer days.

Maybe this was a story of genes and behavior, nature and nurture, Peggy said. The passion for these games certainly came from the boys themselves. "No one *tells* the boys they should play these games."

"Actually," I said, "everyone tells the boys they *shouldn't* play these games."

"You keep wondering what skills they're learning that will be 'useful in later life.' "

When we had cleaned up, John Heywood and I went into the Heywoods' den. He was usually so quiet that I wanted to talk with him alone to get to know him a little better. In the Heywood family, Peggy talked and supported and understood. John did not talk as much, but he also supported, and his sons felt he also understood.

John belongs to a singing group called The Grace Notes. It was started by a former teacher of Jamie's. The leader of the group seems to think of John as the very model of the Dignified Englishman, so he is always casting him against type—making him "Honey Bun" from

South Pacific in a coconut bra. That month John was rehearsing "Well, Did You Evah" by Cole Porter whenever he was stuck at stoplights and airports.

What a swell party, a swelligent, elegant party this is!

John spoke with a deliberate distance, an effort to survey from a height, to speak with composure and long perspective. A British cool that was not frosty. He told me how much he valued the closeness of his sons and his family—I think that was what mattered to him, not the nature of Jamie's project but that he was doing it, that Stephen had come home when he was sick and that Jamie had come home to try to save him. John had not grown up in such a family himself. "My family is small. Peggy's one of five. And they all have kids. And they all get together. Very different from mine. And, *any excuse.* Well, not just any excuse, but they get together *often.* I've learned a lot about extended family from her. And our kids have picked that up, too."

When I asked John what he thought about Jamie's chances with his gene therapy, I was surprised to hear how pessimistic he was. That was an even bigger surprise than when I had asked Matt During the same question. "You're not going to hit a home run the first time," John said. "In fact, it's unwise to try." He served on Jamie's board of advisers and went to the weekly meetings. But he let me understand that he stayed out of it for the most part because this was Jamie's project. His role in the family was to be the provider. "My job is to do my job so there's a good income coming in." They had to be saving for later, John said, staring down at his hands, his palms cupped upward. Stephen's illness would soon be costing them almost $200,000 a year.

While we were talking, the phone began ringing and ringing in the kitchen, and we ignored it. Peggy had already gone to bed. The phone kept ringing. A few rings, then stop, then a few rings again. The answering machine was turned on, and I thought I could hear shouting from each message. Finally I told John that we should answer the phone. When he played back the tape in the kitchen, we heard Jamie's voice. He was calling from the Victorian. He sounded elated, almost manic. "Pick up the phone! Dad, pick up the phone!"

When we phoned Jamie, he was still shouting. John put the call on speakerphone so we could both hear Jamie's voice: "Jeff Rothstein just called me. He met with Marge Sutherland. *The mouse is still alive!* It is so far beyond any record that Jeff is jumping up and down. He's off the charts."

Stephen got on the line, too. He shouted, "The mouse that should be dead is still alive!"

Margaret Sutherland was the young epilepsy researcher at Vanderbilt. Rothstein had just bumped into her at a neurosciences meeting and asked her how her work was going. And she had told him. And then Jeff had called Jamie. They had both been waiting all year to hear about her experiment. Sutherland told Rothstein that she had crossed just one ALS mouse with one EAAT2 mouse. In that litter, only a single mouse had inherited both the ALS and EAAT2 genes. That mouse had gotten both the poison and what Jamie hoped was the antidote. All of the other mice in the litter had died as expected, at the age of five months, paralyzed and choking.

The mice were born in January of 1999. All but the one with the antidote was dead. Now it was September and that mouse was still alive.

"Dad, drive Jon over here!" Jamie said.

As we drove through the dark suburban streets, John was still just as quietly and sturdily skeptical as he had been back in the den. His son was not a biologist, he said, and these mouse models were notoriously unreliable predictors of human cures.

In a minute we had parked outside the Victorian. Inside we could hear Jamie, Stephen, and Robert yelling.

This was the property that Peggy had bought the boys while John was out of the country. She had bought the boys the house knowing and trusting that he would support her completely. I asked John how he had felt when he got home and saw the place. He gave me a wry tilt of his head, and a shrug. He had accepted it. I loved that gesture in the suburban dark, part smile, part grimace, part shrug, man to man, reticent and candid. It said more than a speech.

The ground floor of the old house was full of shouting and maniacal laughter. When we walked in, we saw Jamie and Robert each sitting at a desk and staring at the screen of a computer and jiggling a joystick. Stephen was off in his computer room with the door open. Jamie had wired all the computers in the house into a network and the three of them were playing Starcraft.

"So that's exciting," John said quietly.

"That's *very* exciting," Jamie said. "Dad! It's totally awesome!"

John gave me a level look across the room. "One mouse," he said.

"One mouse," said Jamie. "The oldest mouse in the history of ALS."

John Heywood smiled. He had been through enough experiments and team projects to know that they had many ups and downs. He let Jamie shout and explain. Then he said good night and left. He had to get up before dawn to bicycle to the Sloan Automotive Laboratory.

Since Jamie had called me over, I assumed that he wanted to talk about the mouse. But he was too elated and distracted by the game. The three of them, Jamie, Stephen, and Robert, kept shouting as they played. They looked like my kids.

I wandered through the open door into Stephen's room. Jamie and Robert played with one hand on a joystick, but Stephen's right hand was too weak for that. He had worked out a way to play with a mouse and a keyboard. He was afraid that soon he would not be able to do that. But Jamie told him that if the gaming equipment got too hard for him to use, he would drag Stephen down to MIT's Media Lab and say, *Fix it, guys!*

Stephen tried to explain the game for me while he played, but Jamie took advantage of his distraction and attacked. "So I think what I'm gonna do—no, it's OK, Jamie is clueless—I'm going to—oh, fuck. Jamie just killed me. That's a huge army. That's a big army. OK." He exhaled through his mouth. "I guess at this point it's a bit of a race. Soon it will be a massacre. And one of us will go quickly."

The phone rang. It was Ben, in L.A., returning Stephen's call about the mouse. Stephen told Ben the story. Listening to his answers to Ben's questions, I could tell that Ben already knew all about the background of the Sutherland experiment. Like everyone else in the family, Ben was following Jamie's project closely enough to appreciate what the news might mean.

I wandered over to watch Jamie's monitor. He was so obsessed with his race to save Stephen that TV shows did not engage him anymore. But computer games still worked. He did not leave the battle when I walked up. "I can see this scene is going right into the story," Jamie said, and groaned. But he did not look up from the screen.

Rothstein had picked up another piece of news at the conference, but he had not told Jamie. It was about Jesse Gelsinger, the young man who had entered an OTC gene therapy safety trial in Philadelphia on September 9. It was a Phase 1 clinical trial, a safety trial, to see how big a dose a human body might tolerate comfortably.

Rothstein had spared the Heywoods the story. He let them have a celebration. They would hear about it soon enough. The story was very bad, and the whole world was going to hear about it.

Wilson was using cold virus, adenovirus. Matt During was using a virus he thought was safer, called adeno-associated virus (AAV). Unlike cold virus, AAV in its pure form does not replicate in human beings and does not trigger an inflammatory response. Unfortunately, AAV is also a much smaller virus—a smaller syringe. There is less room in it for the cocktail of DNA.

Those were the trade-offs. AAV was safer but it was so small a syringe that it gave Matt the awkward engineering problem of trying to fit all the essential DNA in, a problem he still had not solved. Instead James Wilson and his group had injected Jesse Gelsinger with a cold virus. A cold virus is bigger, so it is easier to put a DNA package into it; but of course a cold virus is recognized as an enemy by the human immune system, which is why we get rheumy eyes and runny noses and mucus-filled lungs when we are infected by it. In short, a cold virus is efficient but risky.

The primary reviewer of Wilson's proposal, Robert P. Erickson, a geneticist at the University of Arizona, warned the RAC in 1995 that Wilson's plan, delivering the genes directly into the liver through a catheter, might be dangerous because it would expose the liver to such a high dose of viral particles that it could trigger an inflammatory response. He thought it might be safer to inject the virus into the bloodstream through an ordinary IV and let the circulation of the blood bring it to the liver. He also questioned whether it was right to subject relatively healthy volunteers to such a risky trial. But the FDA thought it was safer to inject directly into the liver rather than expose the whole body to the genetically engineered virus.

Wilson and his team had engineered the cold virus to be less infectious than the wild type. It still had that risk but they judged it an acceptable risk and so did the regulators in Washington. There were about 400 gene therapy trials under way at that time and about 100 of them used that cold virus. It had been approved for that purpose by both the FDA and the NIH, and of course the common cold is not usually a killer.

Because the mechanics of gene therapy are so badly understood, the whole effort is horribly inefficient. To get the genes into the patient, gene therapists inject trillions of virus particles, which is a much more massive infection than anyone suffers with an ordinary cold. The therapists inject trillions of viral syringes, and hope that at least a few of them will go on to inject their load of DNA into the right cells and that a few of those injections will make it through to the cell nucleus and insert themselves in the cellular DNA.

Unfortunately, at such high doses of infection the immune system may react vigorously to the attack. That is what had happened in animal studies of the cold virus. Some monkeys given high doses of the virus had suffered such a strong inflammatory response that they had gone into shock and died. A human patient with cystic fibrosis who had volunteered for a gene therapy trial had a toxic reaction, too, though it was not fatal.

Gelsinger was the eighteenth patient in the OTC gene therapy trial. Wilson and his team had already treated seventeen volunteers with their gene therapy, gradually increasing the dose. One of the volunteers had a grade 3 reaction. Blood tests showed that some of that patient's liver enzyme levels were elevated. A grade 3 reaction is significant. By the agreement they had with the FDA, Wilson and his team reported it and stopped the study while they waited for a ruling. The FDA told them to go ahead. A second patient had a grade 3 reaction. Again the team reported it, and the FDA told them to go ahead. When another patient and yet another had a grade 3 reaction, Wilson and his team did not bother to call the FDA. They did include those data in their next formal progress report to the FDA, but they did not highlight those incidents, and the FDA said nothing about them. The team went ahead with the tests, trying to find the point at which the dose was too great, testing the edge.

Wilson was testing three patients with one dose, then increasing it and trying it out on the next set of three. So, Gelsinger got a dose bigger than the last set of volunteers, and much bigger than the first set. In the body, small increments of change can lead suddenly to catastrophically new reactions. The volunteer just before him got the same dose and lived. But Gelsinger already had elevated ammonia levels when he showed up for the test. When he had enrolled in the study three months before, in June, just after his eighteenth birthday, his ammonia level was within acceptable limits, 47 micromoles per liter. On September 12, the day before he was due for his gene injection, his ammonia level was 91. That should have been high enough to stop the test according to the rules set down by the FDA. But Wilson and his team reasoned that the rules said nothing about ammonia levels at the time of the injection, only about levels at the time the patient enrolled in the trial. The team gave Jesse a drug to lower his ammonia level and went ahead.

On September 13, he got a two-hour infusion of the drug into his hepatic artery, which runs straight into the liver. For reasons that are still unclear, the coat of the virus seems to have caused problems in Jesse's body almost immediately. That night his temperature rose to 104.5°. A high fever is very dangerous for someone with OTC. Jesse became jaundiced. His lungs got badly inflamed. He went into acute respiratory distress. One after another, his organs began to shut down. On September 17, the doctors told Paul Gelsinger, Jesse's father, that they could not save his son. The father made the decision to remove his son from life support, and Jesse Gelsinger died.

If Rothstein had shared this news, Jamie, Stephen, and Robert would not have spent Boys' Night Out whooping, hollering, and celebrating in their old, echoey Victorian. There was maniacal laughter from all three rooms. It was like listening to the banter of fighter jocks in a flight simulator. Stephen attacked Robert, and Robert yelled, "Oh, you slut!"

"We're all dead in the end, Rob."

"That's a good point, Stephen."

Rob counterattacked, and now it was Stephen's turn to howl. "Did that seem right to *you,* Rob?"

"How much in this world is truly *right,* Stephen?"

Jamie played in rapt, dead earnest, like one of his science-fiction heroes. It was like watching Luke Skywalker when all the other space soldiers have fallen away and he is racing for that one weak spot in the Death Star, and the whole audience is rooting for him: "Luke, remember the Force! Use the Force, Luke!"

In basketball, when the hero on the court takes an impossibly long shot to win the game just as the buzzer is about to sound, there is often one mad, wishful moment when you are sure, you are convinced, you believe as it reaches the top of its parabolic arc that the ball is going in. The ball is going in!

{ T W E N T Y - N I N E }

Grace Church

"When your quick and dirty experiment comes out positive, it's exciting," John said judiciously, the next morning. He and I were slicing bagels for brunch. John was still steady and solid. "It's only one mouse," he said. "But it's positive stuff that keeps people rolling forward."

Jamie burst in the back door, even more electric than he was the night before. Already that morning he had been on the phone again with Jeff Rothstein in Baltimore.

"That mouse—they thought it was born in January," Jamie said. "It was born in *November!* It's lived seven months longer than it should. This validates the whole idea! Jeff was bouncing. Jeff was just bouncing."

"*What* was the news?" asked Peggy, coming into the kitchen from the solarium, where she had been setting the table.

Jamie explained it all to Peggy. Then he added something else he had not known the night before. "There are some neural complications," he told his mother. "It doesn't learn."

"A stupid mouse?"

"A stupid mouse."

"How do you *tell?*"

"Well, the scientists don't say stupid or smart." Jamie described the mazes that neuroscientists use to test the ability of mice and rats to learn and remember new things. Apparently the mouse, with too little glutamate between its synapses, was not able to learn and remember very well. That made sense. Although too much glutamate can kill nerves, it is, after all, the messenger by which most nerves communicate. If you take away too much of the stuff, the nerves are less able to talk with each other.

Because experimental treatments for neurodegenerative diseases have to treat the brain, they often do risk transforming the minds and personalities of the people they seek to help. The risks of gene therapies for brain diseases are likely to be terrible, as the philosopher Philip Kitcher writes in his book *The Lives to Come.* The operation could be a success, but even so, he writes, "relatives and friends would wonder if their loved one had survived the genetic therapy." Just as we have come to doubt the value of frontal lobotomies and certain kinds of electroshock treatments, Kitcher writes, "we might reject some forms of gene replacement because they failed to preserve—and therefore to cure—the person with whom we began."

I tried not to look up at Stephen as I sliced the bagels. There was no class warfare in the Heywood house as far as I could tell. Stephen was following in the honorable tradition of his mother's father and brother back on the farm in South Dakota. The Heywoods all admired people who knew how to do things, people who were interested in the things of this world. Still, in the American caste system, carpenters ranked lower than engineers and entrepreneurs. Stephen was blue-collar, or trying to be. Jamie and John were white-collar engineers, and until she retired Peggy had been a white-collar psychotherapist. Ben was trying to become a producer in Hollywood. I thought there was something creepy in the notion of the carpenter being saved but rendered stupid by the engineer.

Jamie hurried back to the good news. "This is the oldest mouse ever! It's perfect. It's awesome."

He and his parents were going to church that morning before

brunch, and they invited me to join them. (Stephen never went.) John drove off early to join the choir, and I went with Peggy and Jamie. Peggy wore a floral silk dress. Jamie wore neatly pressed pants and a black turtleneck. He drove, speculating the whole way about the meaning of the mouse that lived. If EAAT2 saved the mouse's life, then glutamate poisoning really must be what had killed all its brothers and sisters. He was right after all! He parked on Eldredge Street in Newton Corner, in front of a house about a block from the church. Stepping out the door of the car, he stood staring at the house. He seemed to expand as he stared at it.

"This is it, Mom. That's what I want."

Jamie was standing in front of a magnificent three- or four-story Victorian mansion. It was yellow with white trim, about a block before the church, just across from a park.

"Why don't you buy that one?" his mother asked drily, but he did not seem to hear her. The mansion was shining in the Indian Summer morning sun.

"Maybe I'll just *make* that one," Jamie said. "And that's just the size I want my porch."

We walked a few steps toward the church and Jamie stopped to admire an even bigger place, a great white Victorian box with a slate roof. A plaque by the front door said, "The Honeywell Club."

As we walked into the cool dark of Grace Church, the choir, in robes, was filing down the aisle, John Heywood among them. Grace Church had marble pillars. Flowers, candles, stained glass windows, a grand ceiling, large crosses. The pews were almost filled.

"It's a Catholic form of service, but no pope," Jamie whispered to me. He was afraid I might feel uncomfortable. "I always used to be embarrassed by how much Jesus Christ there is in the sermons," Jamie said. "They're always talking about Jesus Christ. But even though I hated coming here as a kid, when I come here now, I'm lifted. I love the ceremony, the robes, the songs. I don't believe in a higher power, but I feel something here, even though the words often seem to fight against the meaning and the point."

Peggy tried to put me at ease, too. "Because of the music and ritual I feel at home in Jewish synagogues," she whispered. "There's something evocative and emotional about it. Whatever that sets off in people, it's very powerful. We jokingly say we're faithful unbelievers. But obviously there's more, or we wouldn't be here every Sunday. If people don't have it," Peggy said, "I don't know how they survive."

Back in South Dakota, Peggy had gone to a one-room schoolhouse, where the town's community club met in the evenings. Now she found Grace Church as warm and close as her old schoolhouse. "Here I moved to sophisticated Newton, and I've re-created the old village. That one-room school is Uncle's machine shop now," Peggy reminded Jamie. "He picked it up and moved it."

"Beautiful tin ceiling," said Jamie. He really was crazy about houses. Outside the great stone church I had almost expected him to say, *That's the one I want, Mom!*

The choir sang a processional hymn: "When morning gilds the skies . . ." Then they sang, *Gloria!* The congregation recited together, according to their tradition, the Collect of the Day:

> O God, you declare your almighty power chiefly in showing mercy and pity: Grant us the fullness of your grace, that we, running to obtain your promises, may become partakers of your heavenly treasure. . . .

They listened to the words of the prophet Ezekiel:

> Cast away from you all your transgressions, whereby ye have transgressed: and make you a new heart and a new spirit: for why will ye die, O house of Israel?
>
> For I have no pleasure in the death of him that dieth, saith the Lord God: wherefore turn yourselves, and live ye.

Miriam, the new pastor, gave a sermon about turnarounds, changes of heart. "We all have a fear of the unknown. But we can turn our lives around. There is no one particular way. The key is to figure out what would give our lives more meaning. That's not always easy. But God is right here to help us. The door is always open. A thirteenth-century mystic said you can come where you are. Come again wherever you are. Come again, come."

A congregant read a passage from Philippians. He happened to be a retired doctor, Jamie whispered, and his specialty was ALS.

> Let nothing be done through strife or vainglory; but in lowliness of mind let each esteem others better than themselves. . . .
>
> Wherefore, my beloved, as ye have always obeyed, not as in my presence only, but now much more in my absence, work out your own salvation with fear and trembling.

The choir sang another hymn: "O Spirit of the living God . . ." But suddenly I was full of more feelings than I could stand. I excused myself and hurried outside. It was still a beautiful sunny New England fall morning. I sat in the park across from the church. A blue jay was chuckling. A kid in a white T-shirt walked by on the path kicking a chunk of asphalt, pursuing the diminishing bit. Voices of boys shouted at a game somewhere across the park. Strollers passed me on the path. The trees were all still green, although it was cool enough now for a sweater. The sky was a perfect cloudless blue. I heard the usual rubble of sound of a modern village—an airplane overhead, cars starting, their engines turning over without hesitation. A stout woman in what looked like a blue silk church dress was throwing a stick for her dog.

It felt like a cool summer morning. I walked around the park, past a toddlers' playground and a field where young men and boys, some shirtless and barefoot, were playing soccer. A ref blew a whistle. The first fallen leaf I saw on a side of the path was curling up like Stephen's right hand. I sat down on a bench in the sun. It was too much. I had had enough of the edge of medicine. I thought, I'm going back to the Galápagos.

I went back when the service was over. "All of my children—in spite of unbelieving hearts—love the church," Peggy said as she stepped outside. Stephen's wedding would be held at the church that February. Peggy and I strolled into the churchyard. She and Wendy were trying to figure out the numbers, she said. They were wondering if they could fit everyone in the Great Hall. Peggy told me that John had renovated the hall when he was senior warden. She showed me where Jamie and Melinda had greeted their wedding guests in the churchyard, and where Stephen had made the toast as best man. That year Peggy had been senior warden, and she had taken on the church garden as her project. It was finished and consecrated in May, and the wedding was in June.

"It's a lovely spot," I said.

"It wasn't lovely when we started."

{ THIRTY }

Focus Hocus Pocus

Back at the house on Mill Street, John Heywood had picked up a pizza from Bertucci's to go with the bagels we had sliced that morning. Again we sat in the solarium. Everyone joked that I was getting to be a part of the family.

Stephen, Wendy, Jamie, and Melinda were packing for a drive down to New York and Philadelphia that afternoon. Jamie wanted Stephen to see the laboratory that was preparing his gene therapy; and he wanted all of the scientists there to know Stephen.

Stephen had filled a whole tool chest—a small one—with two days' worth of his GNC vitamins and ibuprofen. Too many pills, he said. He tried to take it all on a full stomach and not with coffee.

"That's hard to do around here," I said. "That means, while you're asleep." The middle of the night was the only time the Heywoods were not drinking their bottomless cups of coffee.

I was beginning to feel like an honorary Heywood, and Jamie must have heard a special note in my voice. He studied my face. "He's hooked," he announced.

"No, I'm not!" I said. "I'm going back to the Galápagos." But he was right. I was hooked, and Jamie knew it.

Stephen grunted as he got up from his chair for a second bagel. He stumbled and almost tripped. "Sorry," he said. Whenever he stumbled now he always apologized with a sudden guilty expression. The stumbling did not bother him, but he knew that it horrified his mother and his brother. He was going back to his packing. "Are you gonna bring a razor?" he asked Jamie.

Jamie refused to answer him.

"Bring a razor, Stephen!" Peggy said.

"Bring a *toothbrush,* Stephen," Jamie said.

"Quickly—the mouse story," John Heywood said. He asked Jamie to tell the story one more time. So Jamie explained it all once again. "The mouse is still alive and kicking and happy. It's lived longer than any ALS mouse in history. Jeff Rothstein now knows that this is the most therapeutic treatment in ALS, period, ever."

Peggy analyzed the situation. Now, she said, Matt During and his postdoc Dave Paulsen would have to speed up and inject that EAAT2 gene into adult ALS mice and see if they lived longer.

"You want to crochet an ALS pillow for Dave?" Jamie asked her. "I'm totally psyched."

"I still think the priority is to get those results re-created," said Stephen.

"Jeff Rothstein's doing that at Hopkins." Jamie told Stephen that he had already promised to send money to Rothstein to help speed the work. He said he had never heard Jeff sound so excited.

"He was pretty psyched when I met him," Peggy said. "He was levitating. I can't imagine him *more* excited."

"No, there's a big difference, Mom," Jamie said. "They spend millions on drugs to add days to the life span of ALS mice. This is seven months!"

"What happened to the rest of the mice?" Stephen asked.

"The rest are all *dead!* They didn't have the gene!"

Maybe I was wrong, but it was not excitement for Stephen that I heard. It was the exultation of the healer with a potion that thousands were dying for. It was not a bedside or table-side voice. It was the voice of the young hunter almost home with the prize. It was the voice of a man who might soon be able to buy the Honeywell Club.

I was happy that Jamie was excited, but I wished I heard less vainglory and more fear and trembling.

Stephen stared straight ahead, poker-faced. He held his right arm cradled protectively in his left, and in his left hand he held one of their mother's coffee mugs, with the words "I" and "You" and a big red heart between them. He took a sip of coffee.

Jamie was accelerating into warp drive. He would build his foundation on the cure of ALS and go on to cure orphan disease after disease. He told his parents about an idea of Robert Bonazoli's, setting up a meeting for top pharmaceutical executives and quizzing them: "You did these other orphan diseases in the past. What made the difference there? What can we do to get you to work on ALS? Or the next orphan disease, or the next?"

"That's a good idea," John said.

"It's not on my critical path."

"Maybe it should be."

"Don't distract me."

Peggy laughed. "Back in the days when they were working in the basement," she told me, "they had a sign painted on a board down there, 'Focus Focus Focus.' A friend changed it to 'Focus Hocus Pocus,' and Jamie didn't notice for a long time."

"A friend who's not able to be as gung ho as the others," Stephen added, in a mordant murmur. "To let go of his *doubts.*"

He was a great one for one-liners. Wendy called them Stephen's zingers. I wondered if he was having doubts of his own. I felt afraid for him.

Outside, above the glass ceiling of the solarium, the leaves in the Heywoods' trees were all still green. They looked as if they might last for months. Stephen himself looked as if he were in the summer of his life, but there was a pallor in his cheek under the black stubble, a hint of winter, a suggestion that, sooner for him than for the rest of us, this season would have to turn.

Jamie went right on talking. He told us that ALS patients sometimes buy ALS mice. "They're $250 each. Some patients keep them in these little Havahart traps. They feed them vitamins, they feed them bee pollen, they pray for them. And the mice always die at the age of five months. They're so desperate, these ALS patients!" Jamie said, as if he had forgotten who was sitting next to him. "They've just got *nothing!*"

{ THIRTY-ONE }

Profit

Melinda and Wendy were driving down to New York in one car, and Stephen and Jamie in another. I joined Stephen and Jamie. I thought I would lie down in the backseat and rest while Jamie drove, but he was still wound up. He said there was something he and I had to talk about right away. He ordered Stephen to get in the backseat.

Jamie pulled onto Route 95 and began to speed south, switching lanes often. By way of preamble he asked if I had read *Into Thin Air* or *A Perfect Storm.* He suggested that I think of those books as a model for my article. *A Perfect Storm* had just the right sense of building opportunity, pressure, tension, and moral crisis.

Now that I was hooked, I noticed, Jamie was beginning to direct me, along with the scientists and their teams. He was all over the story, full of angles and ideas, and he was hard to keep out of it. He kept trying to keep me hooked and motivated.

Jon, I know you want to be objective and detached, but you will make a difference, Jon. Your story will make a difference.

Lines like that are daily fare for most journalists, but they were new to me. The scientists I wrote about had never had their portraits painted and did not want to sit for them. They just let me watch them work. Jamie not only wanted a portrait, he wanted to hold the brush.

"Of course," I said, "in the end of *A Perfect Storm,* everyone on the boat drowns."

Silence from the backseat.

Next Jamie asked me to go over the rules for "off the record" and "on the record." He had a speech to make but he might not want any of it to go into the article. For now he asked to keep it off the record.

I did not write any of it down, and I will leave the speech off the record, but not the fact that Jamie made it—and made it with an energy that seemed to feed on itself and build hour by hour. He interrupted himself only to argue the route with Stephen in the backseat. Between them, the Heywood brothers seemed to know all the roads between Newton and New York and they sparred at each fork.

Jamie's speech was about profit. Here, of course, Jamie was wrestling with himself. It was nothing new, only a more passionate and highly caffeinated version of a speech he had been making for some time. He used to call me and anguish over whether he should be setting up a nonprofit organization or a for-profit start-up. I wondered for the hundredth time how much it mattered if Jamie was profit or nonprofit. He was upping the ante, making this a race for a fortune as well as for a life. But after all, I thought, Stephen must have had something like this in mind from the beginning. When he gave Jamie those checks, he had said, "We're going to need money."

"I guess Stephen was your first investor," I said.

"Yes, I guess he was," Jamie said.

I talked with many people about the money question that fall, trying to get it in perspective. Harold Shapiro, the economist, then head of the National Bioethics Commission and president of Princeton, spoke to me several times about the Heywoods in his office on campus.

"Gosh, after all, he's not conscripting people," Shapiro said, when I told him the caliber of the scientists who were working with Jamie. "Nobody's ordering all these other people to do this. They must find something very attractive here. And the kind of people we're talking about, it's not only money. Right? Because they have lots of ways to get money."

Shapiro's commission had recently looked at the question of patient consent. I asked him, "This young man has only a short time to live without these therapies. Can he give true consent in a situation this desperate?"

Shapiro said he could not judge without talking with Stephen. "But I've known people, unfortunately, who had ALS, and they were perfectly competent to make decisions about their care until very near the end. So I think it's perfectly plausible that this person may be quite capable of giving consent in the normal sense of the word. He's very different from the prisoner going to the guillotine, so to speak. He is a free person. The state hasn't robbed him of his autonomy.

"Still, he's clearly vulnerable. This is a very complicated case for a number of reasons, and one is that the procedures are all experimental. So even if someone is extremely well informed, the probabilities are very hard to assess.

"But I think I put a lot of stress in my own mind on people's autonomy, their capacity to live their lives the way they want to."

What about the money?

"The fact that people make money out of something doesn't give me a problem," Shapiro said. "My view is that most people who make money out of things do so because they've done something good. Now, if these people make a billion dollars because they cure ALS, I will give a big huge cheer and be very glad they have a billion dollars. I'll have a little less myself, and so what. If I ever get ALS," he said, with a chuckle, "I'll have a cure."

Shapiro was all for Jamie and Matt as long as they did not cut corners, and got proper approval—and as long as Stephen gave his informed consent.

"He seems quite lucid and reasonable about all of it."

"I could well imagine he is," Shapiro said with feeling. "And I wouldn't want to try in some patronizing way to deprive him of the right to use his life in the way that seems most sensible to him, given his situation. I certainly wouldn't want to be the one to tell him how to behave."

On the other hand, one ALS specialist I talked with was enthusiastic about Jamie's race until I mentioned the money. Then the tone of the conversation changed completely.

"I think it's a dicey thing. That wouldn't have been my goal. Seems a little sullied. ALS is such a difficult, sickening thing. And to make money off of it. Oh my God! Oy gevalt! As long as it's all plowed back into the foundation—as long as Jamie doesn't benefit, and it's all to go to Stephen, it's OK, because Stephen may not be employable. I don't see a problem with that. But for the whole family to profit! I don't think I'd want to make money off my brother's misery. Would you?"

I also talked to Art Caplan, the bioethicist at the University of Pennsylvania, about Jamie's dilemma. "Makes me very queasy," Caplan said. "The motive is rescue, and rescue is very commendable. But using rescue as a motivation to fund-raise and then turning it to profit is bothersome, because it takes away the moral rationale that many people would have brought to this in the first place. And you'd be fishing in different ponds. You go to the venture capitalist for profits. You go to the church group and friends and the compassionate civic organizations for rescue. You're in different places. And it might actually make people feel they'd been duped."

Caplan thought Jamie had a conflict even without a financial stake in his race. "Should the pace of medical research be determined by people desperately afflicted and their kin? Is that the best way to move the science? My argument would be no. Just as it's hard to do the best science when you're heavily invested and have a financial interest in what's going on, it's very hard to interpret results when your vision is completely clouded up by love of your subject.

"And there are costs involved. The people who give the money may say, 'Do what you can, do the research.' But obviously there's the ethical question of how to handle it responsibly. If you throw the money out the window on science that isn't ready to go, then you haven't necessarily done the best with the gift.

"Medicine doesn't normally move by trying to do something desperate out of compassion.

"How'd you hook up to these guys?"

James Wilson, the gene therapist, had stock in the biotech company Genovo and so did the University of Pennsylvania. Wilson owned a large stake: thirty percent of Genovo. One year later, Genovo would be bought by Targeted Genetics Corp. of Seattle. (Their Web site says: "We discover, develop and deliver molecular medicines to cure disease.") Wilson would get 13.5 million dollars in Targeted stock, and his university would get about a million and a half.

Jamie was struggling with not one but two conflicts of interest—one with the gene therapy and one with the stem cells. They both felt to Jamie like hairy conflicts. (The neurovaccine he was setting aside for the moment—it looked too dangerous to try.)

With the EAAT2 gene therapy, since Jamie was involved with a nonprofit organization, since old women in Grace Church were pressing their tithe money into his hand on Sundays, he should not be able to make money. But what if doing it for profit would make the work go faster?

The stem cell project that he and Matt had started was different. That was for CNScience. They talked about it only surreptitiously around me—they did not want to be scooped by competitors and they did not want me to know things that I might let slip while I made my rounds gathering information for the article. That's why he and Matt had an air of conspiracy when they talked about stem cells—they leaned forward and spoke cryptically and quietly to each other, rather than help me follow the conversation as they did with EAAT2. They were hatching a business scheme.

I learned the details only later. They were excited about a new technique called KDR that might allow them to pick out just the stem cells they wanted, instead of injecting a great slew of candidate cells to increase the chances of injecting the small number of desirable stem cells.

Without KDR they might have to inject 25 million cells in Stephen's brain to get enough stem cells in there to do any good. Injecting that many cells might be dangerous. If they used KDR, they would have to inject only three hundred thousand cells. But it would take time to develop KDR.

With stem cells, Jamie figured that he could make money even though he was involved in a nonprofit foundation, because the foundation had nothing to do with it. That idea came purely out of his brainstorming with Matt for CNScience. If the foundation had nothing to do with stem cells, there was no conflict of interest.

On and on, around and around he went. With the gene therapy, the foundation supported the work. There, both legally and ethically he could not profit, and yet he was sorely tempted to take it to Genzyme and sell it for millions and make it happen fast. In that case, my God, why was he not entitled to make money?

Why not, I thought, if he saved Stephen? Jeff Rothstein had told me what the end is like in ALS. He said most people, even most physicians, have no idea what it is like. It is even worse than the deaths of the fishermen in *A Perfect Storm.* "It's like drowning, only prolonged. People just can't suck in enough air. For some patients it lasts for weeks, for most it's months. Slow lack of air, a constant air hunger. Oh, yeah, it's horrible, it really is. And there's nothing they can be given to be made more comfortable. We give people an external ventilator, but it can't compensate in the long run for the lack of diaphragmatic breathing. There's just not enough strength to bring enough air through your mouth and into your lungs. You're suffocating, your breathing capacity keeps diminishing, your blood chemistry changes, and finally you have a heart attack."

That afternoon Jamie delivered an eloquent lecture in the Socratic style, question and answer, question and answer, and by the end of it I was convinced that he should save Stephen for profit as fast as he possibly could and get rich.

We came into view of Manhattan. It was a beautifully clear, almost hazeless day in early September, one of those days when from many miles around you could see all the way to the towers of the World Trade Center. Did I notice them that day? In my memory, they seem to stand on the horizon like the twin towers of Jamie's dream.

I studied Jamie's face in profile at the wheel, the way I had when we passed Genzyme on the way to his house, the biotech that had built itself on a single orphan disease. I thought that I had never seen anyone so young, gifted, ambitious, and deserving of a fortune. I would make his legend. I wanted to say, *Jamie, if your cure works, and you form a company, I will quit my job and come work for you myself.*

He finished his speech just as we pulled off onto the West Side Highway. The traffic slowed in every lane and he could no longer speed. After a monologue of five hours he seemed just as wired as ever. He kept worrying and gnawing his dilemma.

"A woman comes up to me in church today, and she says, 'I was cleaning out an old purse,'" Jamie told me. "She holds up a twenty-dollar bill. 'Do you take cash?'"

From the backseat Stephen broke a long silence.

"Who?"

It was the first word he had spoken for hours. His voice sounded thin, high, reedy, and small. I wondered if that strange reediness after his long silence was one of the first signs that the ALS had begun to touch his ability to talk.

"What?" Jamie asked sharply.

"Who was it?" Stephen asked again.

"Mrs. Smith."

"How's *Mr.* Smith?"

"Emphysema. A nasty disease. Don't get that."

Suddenly I felt for the hundredth time that Jamie was acting a part, even when he was agonizing. For one horrible moment I even wondered if any of this story was real at all.

Maybe Jamie was sincere in all this manic agonizing; or maybe he was only manipulating me into giving him my blessing. Maybe this was all just a game of bait and switch: The prize was *The New Yorker*, and the fortune that seemed almost in reach.

"Jamie," I blurted out, "are you shitting me?"

"No, I'm not," he said. "Am I, Stephen?"

"No," said that awful voice from the backseat. "He's not."

We met up with Melinda and Wendy at Melinda's father's apartment in Chelsea. Phil Marsh's apartment was large and sprawling, by New York standards, and the wiring work that Jamie had done was still neat, tidy, and rational. Everyone settled down on the sofas in the living room to watch *The Simpsons.* During the first commercial, Stephen said he had forgotten to pack a belt. He asked Jamie to lend him his. He claimed that Jamie had once stolen one of his. Jamie denied that. Phil lent Stephen a belt.

It was time for me to go check into a hotel and get some sleep. I was desperate to be alone again. But Jamie was still wired and he rode the elevator down with me to put in a few more words while I hailed a cab. Out in the street, the sun had set. The cars and trucks were turning on their lights. They were almost bumper to bumper, but they sped almost as fast as they had on Route 95. Jamie stepped down off the curb and raised both his hands. Why was I letting him flag me a cab? I had cast him as a hero, and now he would not stop acting like a hero, even when I wanted to get away. The traffic made Jamie launch full-tilt into a lecture on a new topic: global warming.

I looked at Jamie as he stood there at the curb, young, tall, handsome, elated, as incandescent with power, light, and hubris as he had been that morning by the Honeywell Club. He told me about his father's latest work at the Sloan Automotive Laboratory to reduce greenhouse gas emissions, to help treat the planetary fever. Suddenly I felt tired and mean.

"That's even more heroic," I said.

"He's not trying to cure the problem," said Jamie, and hailed my cab.

{ THIRTY-TWO }

The Sharks I Am Afraid Of

The next morning, we piled into Jamie's car again and got back onto 95, heading for Matt During's laboratory in Philadelphia. Jamie and I sat in the front, Stephen and Wendy in the back. Melinda was staying in New York to spend the day with her father.

As we sped down toward Philadelphia, Jamie talked with the same manic intensity as he had the day before, while Stephen and Wendy murmured to each other in the backseat. Jamie lectured me about the future of artificial intelligence. He talked about computer networks and neural networks at the Neurosciences Institute. He explained his plans for start-ups, his old hopes for turning the institute's ideas into gold.

I told him about a friend I had known back in my student days who had hacked into the Ma Bell telephone system as a freshman at MIT. My friend had built some kind of electronic noisemaker that emitted a series of R2D2 whistles into the mouthpiece of a pay phone and gave him free calls anywhere in the world for as long as he liked.

"And now he's a billionaire," Jamie said.

"Actually, he became a geologist."

"Big mistake."

"No, it wasn't a mistake, Jamie," I said. I was angry again. "He did what he wanted to do, because he loved doing it."

"Ah," Jamie said. "I can hear the wheels turning."

I thought, This guy is way ahead of me. He was also driving eighty-five miles an hour. He changed lanes suddenly.

"Slow down!" Stephen said in his reedy voice.

"You didn't used to be this fucking paranoid before you got sick," Jamie said. He swerved into a new lane.

Stephen protested again.

I craned around to look at him in the backseat.

"I'm glad you're looking out for us."

Stephen had his arm around Wendy. He gave me a look. "It's us I'm looking out for," he said. He said it in a sort of flat black voice, absolutely without intonation. He left it to me to figure out that the word to italicize was *us*. Remember us? Do not lose sight of us.

Long afterward, Jamie and I talked about that trip and the strangeness in Stephen's voice. He had heard it, too. "It was emotion," Jamie said. "I think there was a fair amount of emotion—partly about the visit to Matt's lab, and partly about you. He was feeling like a subject. He felt that I was not taking him along because he was my cool brother but because I wanted him to be a poster child for my project. He didn't relish that role—it was like he was on display. None of us in our family is good at playing that role. I think Stephen was wondering, 'Why am I doing this? What am I adding here? I'm just a symbol.' And Wendy being there. There was a lot going on.

"I never did that again. I didn't put him on display again. Paola, I love her, and Matt is just a joy for me to be around, but they were *my* friends. I don't think Stephen really cared about seeing the lab. What I was doing was bringing up the emotional level for my team. You use the tools you have. Stephen wouldn't say anything, but you heard the emotion in his voice."

In Matt During's office, there was the same picture on the calendar for September of 1999, the sculpture by Rodin. I wondered if it was a present from Jamie from the Louvre, a subtle reminder to Matt, like the ALS pillow that Jamie kept promising to Dave. One of the pictures in the calendar was Rodin's *L'Adieu,* a marble head with white gracile fingers raised to the lips.

Jamie made brisk introductions and got down to business.

"So I have some interesting news, Matt."

He told Matt the story all over again. The mouse was almost ten months old. "Jeff was jumping up and down! He was pumped!"

Matt listened at his desk with his eyebrows crooked upward, tight creases around his mouth and eyes, and scribbled notes. The beautiful screen saver on Matt's computer monitor was a false-color scan of the human hippocampus.

"Good, good," said Matt. "Do you know what promoter she used?"

Now their conversation became technical. It had to do with the additions that go along with a gene to make it work, to make it express itself properly.

Jamie explained to Matt that mice with too many EAAT2 pumps seemed to be perfectly healthy. "Only thing is, they don't learn," he said.

"What do you mean, they don't learn?"

Jamie told him about the maze tests the mice had failed.

"That makes sense," said Matt.

"Yes. If they don't have enough glutamate in the cells, that's a problem."

To Jamie the best part of the news was that it implied he was right: He was interfering with the right molecule in the neural cascade.

Matt listened in his "Thinker" pose, under the Rodin on the calendar. "Makes sense," he said again.

"How far have you gotten to searching down the rights to the stuff?" Jamie asked.

Meanwhile Stephen and Wendy had picked up some reprints of Matt's scientific papers from his coffee table.

Matt's colleague Paola Leone came in to join the meeting.

"I was just reading your paper, or trying to," Stephen told her politely.

Jamie went on talking to Matt, talking strategy now. "Not to get political about this stuff, but to get political . . ."

Wendy wore her big platinum engagement ring. She leafed through a journal, skimming a story with the headline "When Smoke Gets in Your Genes." She and Stephen both looked tense.

I told Wendy softly that Matt and Jamie looked like brothers.

"Only Matt looks more rugged," said Stephen.

"I've lived a harder life," Matt said from his desk as he sketched various DNA cocktails for Jamie on Post-it notes. The sketches looked like little rows of boxcars: promoters up front linked to genes linked to more attachments ending at the caboose of the stop codon. Matt scrawled acronyms above some of the boxcars.

What he was diagramming was a little length of artificial DNA, the piece of DNA they would inject into Stephen's spine. They had figured out what they were going to inject into Stephen now. The hollowed-out viruses they would use would carry very little DNA of their own: All but a few percent would be deleted. So each virus capsule would be essentially an empty syringe. And into each one they would pack that single linear strand of DNA composed of a promoter to make the gene work, and then the gene itself, and then a little more genetic apparatus, including a piece of DNA called woodchuck hepatitis virus postregulatory element.

The woodchuck element was an innovation that Matt had picked up from another genetic engineer, Tom Hope of the Salk Institute. In some tests a fragment of woodchuck element boosted the productivity of the gene he injected by more than fortyfold. Matt had been the first to use the woodchuck element in adeno-associated virus. In one test, he injected the virus loaded with the woodchuck promoter and also a luciferase gene from a firefly into a rat's brain. He could see that the gene got where he aimed it, and that it switched on, because the nerve cells around the tip of the needle lit up and glowed green.

They also had figured out how to get this artificial piece of DNA to the cells that needed it in Stephen's spine. They could concentrate the virus particles at more than 10,000,000,000,000 per microliter in a vial. Injecting three microliters in one spot would be enough to reach nerve cells extending several millimeters all around the tip of the needle. And the surgeon Fred Simeone, chairman of the neurology department at Jefferson, had decided how he would get the needle into the cervical spinal cord. He would go in from the front, through the throat.

Though the treatment they were planning was highly experimental, there were so many desperate ALS patients that Jamie and Matt did not expect any trouble finding twenty-five volunteers between the ages of eighteen and eighty who would be willing and able to give informed consent. They would not even have to advertise; they could recruit them all from Jeff Rothstein's clinic at Johns Hopkins. Rothstein was glad to cooperate, partly because the gene therapy focused on replacing the EAAT2 gene that he had shown is defective in patients with ALS. And of course Stephen would be among the twenty-five volunteers.

Stephen and Wendy sat waiting like patients in any doctor's office. But instead of *National Geographic,* Wendy was now reading a scientific paper from the coffee table with the title "Environmental Enrichment Inhibits Spontaneous Apoptosis, Prevents Seizures and Is Neuroprotective." And one of the patient's doctors and one of his brothers were murmuring with their heads together over a notepad, trying to design a dangerous new therapy for the patient on the spot. Otherwise, it was an ordinary scene.

As usual, Jamie was pressing Matt to do a little more a little faster than he was prepared to do. "Jeff Rothstein doesn't look at behavior that much," said Jamie. "You have a very good behavior lab in New Zealand." He suggested a test that Matt could try, to see if the injections really would make the mice stupid.

"I have a grant to submit in three days," Matt replied in a compressed voice. "It's a huge stress." But Jamie made him promise the test.

As we were leaving the room for Wendy and Stephen's tour of the lab, I stopped to look at the calendar over Matt's desk. Jamie noticed me looking at it and he hung back while Stephen and Wendy got a few steps ahead on the tour with Matt and Paola.

"At the Louvre, I kept looking at Rodin's hands," Jamie said. "I was looking at this stuff and Stephen was tripping on the steps. And I was just gone. I was losing it."

We caught up with the group. Wendy was frowning and following Paola's explanations of the lab's innovations as if her life depended on it. She leaned forward as she listened. "If you're putting this into the brain," she asked, "how are you targeting the brain?"

Paola explained. They injected genes directly into a specific part of the brain, like the hippocampus. She talked about their Canavan trial, and how the virus had been injected directly into the brain to bypass the blood-brain barrier. She was standing in an open doorway marked "Biohazard BL-2." Wendy wore a fixed frown as she listened, which she seemed to erase almost by an act of will whenever she looked up at Stephen's face.

In the prep room, where rows of lab coats hung on pegs, they stopped to look at a white rat in a cage. It was sniffing away at the room as it paced from one wall of its cage to another, its pink nostrils working. The rat was an albino, with red eyes and a pink nose that was sniffing toward them like crazy.

"We used to have a pet rat at Colgate," Stephen said. "Thor."

"What's going on with this one?" Wendy asked Paola.

Paola showed them the laboratory bench where they would operate on the rat. As she began to explain the procedure, Stephen folded his arms across his chest. I remembered how I had felt when I first toured the lab, how charged it all looked with hopeful and horrible possibilities. But when he spoke, Stephen sounded cheerful. "Amazing that all this equipment exists, you know?" he said to Wendy. He was walking more easily, as if he felt more at home with each step, admiring the equipment. "These are great. These are beautiful. Look, on wheels."

Paola showed them incubators where they grew big batches of cells. She opened one to show them the cells growing there by the billions in little petri dishes. The incubator was lined with copper.

"Why copper?" asked Stephen.

"Copper to kill off any stray yeast and bacteria," said Paola.

"See, we could do it," Stephen told Wendy. "Copper counters. I use a lot of copper in my building projects," he told Paola. "It's a beautiful material."

His face kept lighting up at all the neat new Macs in their colorful translucent clamshells on the laboratory benches. Then he got interested in the layout of the new offices. He was concerned about all the doors between rooms and halls and the sections of labs. They made it harder for the doctors and scientists to get around.

Stephen was not afraid of much. But he had now seen Matt's movie about super-smart sharks twice, at Matt's suggestion, and he had developed a little phobia about sharks. In the seminar room, one of the magazines on the conference table was the November issue of *Aqua.* The cover showed a Polynesian woman standing next to an outrigger in blue, rippling shallow water, sharks swimming around her. The cover line read, "Shark Time in Polynesia: Rangiroa's Toothy Attractions."

"Well, honey," Wendy said, teasing him. "It looks like we're not going to Polynesia!"

"This is *not* the kind of shark I'm afraid of," said Stephen, primly.

Jamie found Dave Poulsen, who was preparing the DNA cocktail they would inject into Stephen's spine.

"By the way," Jamie told him, "this is the guy we're doing all this for. Dave, this is Stephen."

"Hi," Dave Poulsen said. He stared down at the floor and then glanced up again. He did not quite meet Stephen's eyes.

It was time for lunch. As we walked toward the revolving doors in the lobby of the building, I was alone for a moment with Stephen because the others had already gone through. Stephen walked slowly. "What do you think of the lab?" I asked him, not very diplomatically, since the next moment we emerged on the sidewalk with Jamie, Matt, Paola, and Wendy waiting there in the sunlight and watching us.

"Good," Stephen said simply. "A little disorganized, but good."

The restaurant that Jamie had chosen was Buddakan, one of the best and trendiest places in the city, with a fountain of water running down a stone wall near the front door, and a big statue of the Buddha inside. Jamie asked Paola to order, and when the food came she dished out our portions from the bowls and trays with happy, festive laughs. But when she thought no one was watching her, I saw her stare at Stephen with a moist film in her eyes, as if she were taking a picture of him to go with her photographs of the Canavan babies. I looked across the table and saw the picture she saw: a tall, dark, and handsome man of thirty, beaming at his fiancée.

Stephen saw us watching him and gave Jamie a slow, careless smile, like a young man with all the time in the world. Then he grinned around the table at the rest of us. Still, whenever he was serious, earnest, asking questions, I saw a kind of pain in his face and eyes. And when he was quiet, in repose, he looked very sick.

When the credit-card chit came, Stephen signed it.

"So you are left-handed?" Paola asked.

He gave one of his mordant laughs. "I am now," he said.

{ THIRTY-THREE }

Bat Wings in Paradise

At the end of the month, Stephen and Wendy moved into their fourth-floor walk-up on Beacon Street. "The stairwell was wide, with long flights of carpeted steps and heavy, dark-stained wooden banisters," Wendy wrote later in a journal and memoir about life with Stephen.

"From the fourth floor, the top floor, you could look down through the gap between the banisters and see all the way to the main landing. It was a tight space, but not so tight that one couldn't fall to his death if he were unlucky enough to lose his balance and go over the banister."

The day they moved into the building, Stephen's footsteps were already heavy, uneven, and alarmingly loud on the long flights of steps. The woman on the third floor cornered Wendy on the landing between the third and fourth floors and gave her a piece of advice.

Keep the noise down if you don't want your neighbors to know your business.

Wendy felt for John when he came to visit, and inspected their two small rooms from the doorway. Even as she opened the door, she could see that he was looking for ways he could lend them a hand. With only two rooms to survey, there was not much John could see to do. There were pictures to hang, of course, but Wendy would hang them herself. They needed a kitchen table, and John offered an old one from the house on Mill Street. Maybe he could help move the couch? "He needed a task and we had none to give him," Wendy writes in her memoir. Stephen was still capable, and the son needed tasks even more than the father. "So John settled for standing in the entryway, rotating his hat in his hands."

Every morning Stephen walked down the four flights and walked six blocks to the train, which he rode to Newton, where he was still renovating John and Peggy's master bathroom. He carried a black shoulder bag across his chest and his right arm swung limply by his side. He used his left foot as an anchor, step-by-step, to lift and drag his right foot along. "In his large body," Wendy writes, "he swayed down the street like an updated, very handsome version of Frankenstein."

The *Washington Post* broke the story of Jesse Gelsinger's death at that time, and in the last days of September it ran on front pages across the country.

The boy's father, Paul Gelsinger, told the *Post,* "I lost a hero." He also said he did not blame the researchers. "They're as hurt as I am. They've promised full disclosure."

He told the *New York Times,* "The doctors are as devastated as I am. It was because of these men that I had my son for eighteen years." He would stand by them "until my dying day."

James Wilson told a reporter for the *Times* that he and his team were conducting an investigation of the cause of death. "We owe it to the field and to the patient, who really wanted to participate and wanted us to learn," he said. At first, Wilson and his team told reporters that they had no warning of trouble. As a reporter for the *Philadelphia Inquirer* put it, "The first and only serious side effect in the trial was Gelsinger's death." Then Wilson said there had been one case of liver damage but that no one in the set of three just before Gelsinger's had any side effects. But the FDA pointed out that one volunteer in the group before that one, a woman named Tish Simon of Union, New Jersey, had shown liver enzyme levels that were five times higher than what is considered healthy—levels that should, according to the rules, have been reported to the FDA. The FDA made much of Wilson's lapses—while saying very little about its own.

Immediately, officials with the RAC, the Recombinant DNA Advisory Committee, began looking into fifty other gene therapy trials around the country that also used the cold virus.

What made the young man's tragedy agonizing was that he himself had not needed the gene therapy. He might have lived a long life with careful diet and fifty pills a day. He had joined the trial purely in a spirit of altruism and adventure.

The University of Pennsylvania stopped the trial. The FDA issued a clinical hold on it. The RAC sent out a warning letter to the laboratories that were using cold virus.

Jamie followed the news through Matt and Paola in Philadelphia. At home in the peeling Victorian he sat down with the paper and held it in front of his face. Day after day, night after night, he could not help cutting her dead—Melinda and most of his old friends, too. He had no time or energy for anything but the saving of Stephen. They could understand that, or not. On November 1, Jamie came home from a brunch with Katherine Evans and told Melinda that he would be paying Katherine $2,500 a week to work full-time writing the FDA protocols for the gene therapy project. Melinda had thought Katherine was out of the picture. She heard Jamie and thought: More money we do not have, and yet another person working at a desk in our house.

She wrote a little later in her journal: "I did not react in an exemplary, supportive & adoring wife fashion—he became very angry & left the room abruptly, I heard a loud bang in the other room which I assumed to be his foot kicking the door but which was his head banging itself into the wall."

She went after him to apologize but they had a terrible fight. Jamie said, *How can you worry about defending your boundaries when they get in the way of saving Stephen's life?*

Melinda scrawled in her journal and wept. "I love Stephen more than anything. I love your family. You absolutely have my support

& always have, every step of the way. . . . I have a perfect right to worry about boundaries & protect my peace of mind—I'm also PREGNANT, by the way, and my desire to have my own private space is doubly compounded by that. Boundaries are closing in on me, dissolving, at every turn—my body's boundaries are shifting—hormones have taken me over—another entity is growing within me—there are 7 people in my house, in my kitchen, around the house on any given day—using my dishes, eating my bread, drinking my tea. . . . I have to stay upstairs with the fan on to block out noise from downstairs. . . . It has grown too big."

And the Jesse Gelsinger scandal grew. On November 21, Wilson flew to Tucson to talk with Jesse's father, Paul, and try to explain his son's death. Wilson traveled there with a reporter, Donald C. Drake, of the *Philadelphia Inquirer,* and Drake described the scene at the close of another front page story about the death.

Wilson had never met the father and he had met the son only once. When he arrived at Gelsinger's small house, he found the man waiting for him outside. Paul was standing next to a dirt bike he had bought that year for Jesse. A reporter had just told Paul one of the first of the nasty revelations in the case: The pharmaceutical house Schering-Plough had seen side effects with cold-virus gene therapy and reported them to the FDA, but the law had required the FDA not to tell Wilson's group about it. As Paul saw it, he and his son's doctors had been wronged.

"He didn't have to die. My kid is dead so they could make billions more," he said. "Love of money is what killed my kid. You guys would have stopped if you knew about what happened."

The father was so wound up that he talked on until dark. At last when he felt he had said what he had to say, he let the doctor go over his theories about the cause of death. The father was still loyal to the doctors.

"You guys are baffled, aren't you?" he said.

By then the papers were reporting that Wilson himself stood to get rich if he could make a success of gene therapy. To Paul Gelsinger that was offensive innuendo. He was still convinced that he and the doctors were on the same side. He asked James Wilson for the truth and Wilson gave him a very soft answer. Yes, he had started a small biotech—many researchers did that these days. But he had only non-voting stock, and he consulted for the company only one day a week, for which he was paid nothing.

Paul wanted to know if Wilson would have made money if the trial had worked. Wilson gave him another soft answer. He talked about the years it takes for a biotech to get a drug through the pipeline. His company, Genovo, still did not have a single product, he said.

That satisfied Paul Gelsinger. The doctor, the handyman, and his son were victims together, he said.

"What you guys were doing was pure."

When I heard the news of Jesse Gelsinger's death, I called Art Caplan, the bioethicist. He had served on the Penn internal review board that had approved the study. Caplan said Jamie and Stephen Heywood should drop their project. "You may have a year or two to live, and some of that time you may think of as just more miserable than you can endure, but you may not want to die *in the morning,*" Caplan said. He knew and respected Matt During, he said, but he doubted Matt's ALS project would be approved by Jefferson's internal review board, much less by the RAC or the FDA. "Everybody's gun-shy about gene therapy now," Caplan said.

So was Caplan. "I don't think gene therapy is quite entrails and, you know, bat wings, and the inner gonadal parts of monkeys. But closer to that," he said, "than it is to penicillin. We're still so early on. I understand the desire to want to take charge and do something and pursue something, but racing to find the first cure from gene therapy for a complicated disease is really—that's not just long odds, it's google-number long odds."

As I listened to Caplan, I felt vindicated in my fears for Stephen; but at the same time I felt furious at Caplan for judging the Heywoods without even knowing them. I felt glad that I had not put my mother under the needle and the knife; but I felt outraged that the world had made these new medical experiments so charged that they had a supernatural aura, and now a single failure in a clinical trial could slow or stop them all. I wanted Caplan to tell me the right thing to do, but at that moment I hated him for sounding as if he believed that he could. As a professional I admired his confidence. But it was like calling a talk show psychologist. How could he shoot them down when he did not even know them?

"Well," I said, "if you were talking to Jamie Heywood, what would you say? Would you try to talk the family out of doing this?"

"Yep. And by the way, I can give this speech, because I've given it. Not quite as coldly as this, but I've given it. I'd say the best thing you can do for your brother is try and prepare"—he paused—"for the worst outcome. To line up appropriate spiritual support, and psychological support, and do everything possible to prepare for, ah, death."

When I told Jamie what Caplan had said, he was so enraged that my pen stopped and I did not write anything down.

"What does he think of us? Does he think my family is stupid? Does he think we don't know what is happening here? Prepare for death!" Jamie cried, in a voice thick with rage and tears. *"What does he think we're doing?"*

Later, he said, "It's a good thing Stephen is not Art's brother."

With the Gelsinger scandal, Jamie's job became much harder. He panicked. It was the same terror that he had felt at the very start of his race, he said, "just sort of sheer panic as you try to describe something into existence again." He was still meeting with potential donors and investors and he was still trying to decide whether to go profit or nonprofit, but every conversation was harder. "Sometimes you wonder what great wealth gives you, and I think maybe it gives you the ability to skip this part," Jamie said. "You take on a lot of responsibility, and you know how shaky everything you are describing is, because you are the only one who knows where the strong parts are and where the weak parts are. It is sort of like pure force of will."

The safety record of gene therapy was still far better than the record of most other kinds of drug development, he said. Its efficacy rate was truly abysmal, but its safety record was good.

"I don't know if anyone realizes quite how hard this is to do," Jamie told me. "It was much easier for me to talk about putting together Internet companies. This time it is just much *more.* I wonder if people are always this scared. I met with this CEO—I didn't get the sense that *he* was that scared.

"There is also this sense of being an impostor. 'What the hell are you ever doing here, Jamie, there's fifty brilliant doctors trying to cure this disease, what are you doing here?' I mean, I know I can answer that question. But if Jeff Rothstein screws up, there is no loss for him, he's still a professor at Hopkins and he works hard. If I screw up—I feel like I'm walking twenty-four hours on a plank."

On Friday, December 3, after working a few back-to-back twenty-four-hour days, Matt During, Paola Leone, and Jamie Heywood submitted a proposal for a clinical trial for ALS gene therapy to the internal review board of Jefferson Medical College. They asked the review board to approve their trial in spite of the scandal. In reaction to the death in Philadelphia, the FDA had ruled that it would not review a gene therapy protocol without its first being submitted to the RAC, and the RAC would not review a protocol unless it had been approved by an internal review board, an IRB.

IRBs were mandated in 1974 by federal law after scandals in which doctors had badly abused patients in the name of science. The most notorious case was the Tuskegee syphilis study. There, doctors in Alabama deliberately did not treat black men so that they could watch the syphilis progress until it killed them. The disease spread to some of the men's wives and children, and still the doctors let the study go on.

The institution of the IRB is clearly needed; but the board members are usually unpaid and badly overstretched. One government report found that in a typical university medical center IRB, the board went through agenda items at a rate of less than two minutes per item for two and a half hours. Now Matt During asked his university's IRB to hurry because the next RAC submission deadline was January 13, 2000. If they missed that deadline, the next would come three months later. Jamie and Matt would lose those three months.

On December 8, the fifteen members of a RAC panel, mostly scientists and ethicists, met for a three-day session. The meeting was a sad circus of news conferences, reporters and scientists and cameras, and representatives of government and the biotechnology industry. Wilson spoke to reporters in the company of a press aide and a lawyer. He and the others on his team said the problems with their study came down to technicalities and judgment calls. W. French Anderson, the doctor who is usually called the father of gene therapy, told a reporter for the *Times* that in a big laboratory like Wilson's, "sometimes things fall through the cracks." Jesse Gelsinger's father sat in the back of the hearing room. He was still supporting the doctors: "These guys didn't do anything wrong."

Meanwhile, on December 9 Matt met with a small subcommittee of Jefferson's IRB to present their case. He pointed out that unlike the notorious trial at the University of Pennsylvania, they were proposing to treat people with a fatal, incurable, rapidly progressive disease. Their virus, unlike the virus used in the Gelsinger tragedy, had been shown again and again to be safe. It had been used, for instance, in genetic therapy trials for cystic fibrosis and hemophilia. In those trials there were hints that gene therapy might be working at last.

The board would not reply for a month. There was nothing to do but wait. I called Steve Gullans, a professor at Harvard Medical School who knew and admired Jamie, and I asked him what he thought the IRB should decide. "This is a tough call, a judgment call," Gullans said. "No doubt breakthroughs are rare, and they are all preceded by naysayers saying it will never work. But with gene therapy, you have two one-in-a-hundred shots. First that the virus will get the gene in, and second that the gene will help. If there were just one long shot, I'd be happier. But two—"

Jamie was nervous waiting and he was also nervous about the *New Yorker* story. He called me almost every day. He had found a publicist in Boston who spent a weekend with the Heywood family for free, coaching them on how to make the best of the press attention that would follow the article. "You're not going to let me see it, are you," he said one day. "I asked for this, but this is scary. This is very scary. *Should* I be scared?"

I said he had nothing to worry about, but he did. The money question hung between us now. I was so worried about it that one day I called Ralph Greenspan in La Jolla to ask if he thought I should abandon the story. It was a Sunday afternoon, and we talked for an hour while he did his laundry. Ralph said my worries mystified him. Jamie was an entrepreneur. He was trying to save his brother with all the tools of the entrepreneur. What was the problem?

But I was hugely relieved a few days later when Jamie called to tell me that he had decided to donate all the proceeds of his gene therapy patent and project to the foundation. He had decided at last, he said, in a strained voice. Anything that came from it would belong to the nonprofit, just as he had pledged in his speeches to Robert, Melinda, and Peggy down in the basement.

By now, Jamie could not talk or think about anything but the gene therapy. At home Melinda went on writing miserably in her journal. "Boundaries are the central issue," she wrote on December 19. "It is difficult & I want to flee the scene. I could bear it all if . . . if . . . if only." *If* she had privacy. Or *if* she could hope to have privacy someday. But the baby was coming and they were restoring the carriage house next door for Stephen and Wendy, and the race would go on. "And the *if only* is Stephen—if only we could be sure to cure him & his disease. Instead of feeling that I am growing a life within, I feel I am dying too. We are all dying as this happens. I can't stand it. Hemmed in. No way out. Marriage must be preserved, but my vigilance is crucial."

They dug two trenches, his and hers. Melinda's line was, *What about me? What about the baby? What about us?* Jamie's line was, *You're making it harder for me to save Stephen!*

To Jamie, the gene therapy was the best and almost the only hope, though he and Matt were racing now with other treatments, too. They talked more and more about stem cell injections. Jamie was not sure he wanted Stephen to be the first ALS patient to try that procedure. And as it happened, another patient who was following Jamie's race liked the promise of the idea and decided to try it with his doctor. It was Steve Fowler, the L.A. musician whom Jamie and Stephen had met at the very start of their adventure. By now, he and Jamie were friends. Steve Fowler's doctor injected one hundred million cells into his spinal fluid over a period of two days. On the second day, Fowler lost sensation in his lower limbs. But after two hours the feeling came back. He noticed no other symptoms, and they all waited to see what would happen.

Stephen was weaker now. He and Wendy loved Back Bay, and they were very happy together. Wendy got pregnant. But as they furnished their apartment, the stairs were a strain. The woman on the floor below them complained that she could hear Stephen as he thumped across the bare wooden floors. His footsteps sounded heavy and dragging. She demanded that Stephen and Wendy buy rugs. They went out and bought enough eight-by-ten-foot carpets to cover almost all of their floorboards. As Wendy lugged the first carpet roll up the stairs, Stephen limped behind her, empty-handed and embarrassed. Their neighbor on the first floor smirked at Wendy.

You seem to have the short end of the stick.

When I called one morning to check in—"What's new?"—Stephen apologized for his mood. He was a little grumpy, he said. Their neighbor one floor down was still complaining. That morning he had tired himself out trying to cut mats for the rugs.

Suddenly his breathing sounded very labored.

"What are you doing now?" I asked anxiously. "Still working on the mats?"

"Ah, no, no, no," Stephen said with a weak laugh. "No, I'm just rolling over."

He told me that he felt more and more willing to do something experimental, and sooner rather than later. He had begun telling Jamie, *We gotta get going!* He was the one who was racing now, he said, and he laughed weakly again.

"So, that's new."

My calls to Providence were just as alarming. My mother found it harder to walk, talk, or handle the telephone. She often preferred Yiddish now, the language of her parents in Brooklyn when she was a girl. My father would hold the phone to her ear, I would ask her how she was, and then the three of us would wait. We were all happy if she managed to say, "*A miesele, a meisele.*" So-so.

Early in December, I heard back at last from her neurologist, Stephen Salloway. He wrote a brief e-mail to say that he would get in touch with Matt During.

An editor at *Time* asked me to write something for a special millennium issue, "After 2000." He sent me a long list of essay questions, including "Will We Run Out of Gas?" and "How Hot Will It Get?" with one question circled for me: "Do I Have to Grow Old?"

I called my mother. I thought she might be amused. "What do you think of that? 'Do I Have to Grow Old!' "

There was a long pause. Then, in a Yiddish-like singsong, my mother replied, "I wouldn't recommend it."

One of Robert Bonazoli's sisters made an emblem for the foundation. Jamie and Melinda sent me one with a note of thanks for the work I was doing. It was a long, old-fashioned golden key with an angel's wings, "The Key of Hope." My mother would have called it a *chochka.* I kept it on my desk.

At the Heywoods' second fund-raiser, a Christmas concert in Boston's Jordan Hall, "A Concert in the Key of Hope," Stephen told the crowd that since his diagnosis he had met many ALS patients in various stages of the disease. "No matter what stage you're in," he said, "you always think, 'As long as it doesn't get any worse I can handle this'. . . . I think the key is hope. I know my own progression has been slower because I have hopes for a cure in my lifetime.

"Remember that music is good medicine. So sit back, open wide, and enjoy the medicine."

As the crisis advanced, Stephen got closer and more solicitous of Wendy, while Jamie and Melinda drew further apart. Melinda tried hard to be understanding, but one night in late October she dreamed that she had kissed Jerry Seinfeld. She confessed the dream to her journal: "He had refreshing peppermint breath & a warm moist mouth that was lovely to explore. . . ."

A few nights later, after her big fight with Jamie, she had a dream involving Adam Sandler. He wore a thin, well-worn, sweaty cotton T-shirt. "We spent the night cuddling chastely on his bed. Adam stroked my hair and held me humidly, and I felt like I was back in high school again."

Late in December, she had a third dream. She thought about it all morning at her desk at the foundation. This time she had sex with Jon Stewart, the host of Comedy Central's *The Daily Show*. That morning at her desk, while she recorded donations, sorted mail, and wrote notes, she realized what the dreams meant. "My waking life has become so serious, so tragic, that I must resort to sleeping with comics in my dreams."

At my own desk that December, I was farther away from the tension. But I was drafting the article for *The New Yorker* now, and I was calling Providence more often than Providence wanted me to call. I kept picking up that golden key with the angel's wings and staring at it. I felt oppressed, depressed, overwhelmed. I wanted to escape. As the year ended, I went back to the Galápagos, this time with my wife and our boys. We saw in the new millennium there.

Part Five

The Sudden Fall

The whole house seemes in whirling

motion,

The fixed columns turning round,

and all

The turning roof threatens a sudden

fall.

Lucretius

{ THIRTY-FOUR }

Fires

In the first week of January 2000, I arrived at Rockefeller as writer in residence. That same week ninety-nine years before, the first grandchild of John D. Rockefeller had died of scarlet fever. Now, of course, the place was one of the capitals of an empire. Rockefeller had given scientists a place to stand, and they had moved the world.

My only official duty was to teach a seminar to the young biologists and medical students, "Parallel Lines: Science and Literature." The students were working at their laboratory benches for twelve or fifteen hours a day. I was there to give them an excuse to read.

It felt strange to me at that moment of crisis, when things looked bad for the Heywoods and for my parents in Providence, to arrive at the point of origin of the Neurosciences Institute and so much of biomedicine. Rockefeller was in a celebratory mood that year because it was approaching its centennial, and hundreds of millions of dollars in donations were pouring in from the descendants of the Rockefellers and Astors who founded it.

It was very strange to arrive at the point of origin of so much hope in the middle of private despair, and like Jamie I searched for hopeful signs.

Standing at the window of my office on one of my first days there, I happened to look out exactly at noon. A tall man with a ruddy face and a head of thick white hair was strolling down the esplanade from the northern gate on 68th Street. A few elegant dogs with long pale fur pranced around him. Just below my window, which overlooked the center of the esplanade, he paused for a moment, leaning forward. I saw him stare straight ahead as if he were taking in the long alley of the old sycamores and the laboratories. He looked as if he were receiving some private message about the way behind and the way ahead. He straightened his back, raised his head, expanded his chest, and his face flushed. He wore a dark, dashing coat—a garment with a foreign accent. Somehow even his pose had a foreign accent. He stood there as if the rows of sycamores that framed the esplanade had suddenly turned into triumphal arches, and the stone columns, the ivy, the laboratory windows and hospital windows around him had just delivered the old imperial salute: *Live Forever!*

I knew it had to be Günter Blobel. At that moment when I saw him below my window, he could have been elected mayor of Dresden. Everyone in Dresden knew him on sight: the tall man with the great mass of white hair like the head of a Corinthian capital. He was a hero who against incredible odds had tried to heal a hurt as nearly cosmic as we can imagine on this earth. He was sending a signal of a kind, and it had an effect in the tortured city he had seen in its last days of glory as a boy. He had begun shuttling back and forth, raising money and helping to plan the regeneration of Dresden. He was helping the city rebuild itself, and better, to rebuild its image of itself from a symbol of absolute evil and total loss to a symbol of hope, the recovery of hope in the face of despair.

During the Heywoods' crisis and my family's crisis I did a lot of standing at my office window. Every day at noon the man would stroll down the esplanade with his prancing dogs. Every day he would pause under my window and stare ahead with that same look of more than mortal pleasure. For me he seemed like a figure from the old romance of pure science—for me and for many people at other windows along the esplanade. He had to know that he was being watched, but he did not seem to mind, and neither did he seem vain in his parade. Well, maybe a little vain. But somehow he moved as if, as private as his moment was, we were all involved in the procession. He had kept the faith, and anyone looking out and seeing him there would keep the faith with him.

My desk was a section of a big *L* that had been sawn apart. It was hideous and filled most of the room. But in front of the window there was an old writing desk with gold leaves of ivy around the cracked leather top, and brass clawed feet on wheels. I placed a little edition of Lucretius ceremonially in the center of the desk. There was also a big leather sofa with brass nail heads. When the intellectual-property lawyer stopped by from down the hall, she said, "Oh, that's my sofa." She had donated it to Rockefeller. She also knew Jamie Heywood. She had talked with him about the patent on the EAAT2 gene. She asked me, "How are those boys doing?"

Downstairs in the president's office, I felt subversive when I talked to Arnie Levine. I knew Arnie from Princeton—he had built and run their department of molecular biology. When Arnie left Princeton to run Rockefeller, he invited me to come. Arnie knew what I was writing about. He told me that he despised the hype and flimflam around gene therapy and regenerative medicine. I once asked him about William Haseltine of Human Genome Sciences ("Hype artist . . ."). Sometimes I thought that when Arnie read my story, I would be out the gate.

On January 13, 2000, the institutional review board at Jefferson voted to support Jamie's gene therapy protocol. If the protocol was approved that spring by the RAC and the FDA, the first trial could begin in June. But on January 21, the FDA announced that it was shutting down all of the University of Pennsylvania's gene therapy trials because of eighteen procedural irregularities in the trial that killed Jesse Gelsinger. The FDA stopped eight trials, experimental treatments for cystic fibrosis, muscular dystrophy, lung cancer, skin cancer, breast cancer, and brain cancer. The university began an internal inquiry.

Biotechs put their gene therapy trials on hold. Some universities canceled theirs. So did the Muscular Dystrophy Association and the Cystic Fibrosis Foundation. In early February, the Senate subcommittee on public health held a standing-room-only hearing on gene therapy. "The oversight system is failing us," said LeRoy Walters, director of the Kennedy Institute of Ethics at Georgetown University, and former chairman of the RAC. The gene therapists were supposed to be reporting problems to Washington and they were not. There had been almost seven hundred incidents, known in the RAC's jargon as "adverse events," in gene therapy trials. Some of them were serious side effects and some of them were seizures and fevers that dying patients might have suffered anyway. But only thirty-nine of these incidents had been reported immediately to the NIH. Most had been reported only after the death of Jesse Gelsinger.

"I think the oversight system is failing to prevent serious violations of patient protection," said Senator Bill Frist, who was convening a hearing into the death. "Let's have full transparency in the system."

Jesse Gelsinger's father, Paul, flew to Washington again for that meeting. He was following the proceedings closely, and he was beginning to realize that he had been as green and innocent as his son. By now, he said, trying to keep control of his voice, he felt that he had been misled. "Looking back, I can see that I was fairly naive to have been as trusting as I was." He was shocked and badly hurt. "The concern should not be on getting to the finish line first, but on making sure no unnecessary risks are taken, no lives filled with potential and promise are lost forever, no more fathers lose their sons."

Paul Gelsinger sued, and the University of Pennsylvania settled with him for an amount that was not disclosed. Insiders guessed it was about ten million dollars.

Everyone's gone crazy," Matt During said at Jefferson. He was getting hundreds of calls and e-mails from colleagues, reporters, and patients. Some of his patients, especially the parents of Canavan babies, were terrified because they thought the FDA was stopping all gene therapy trials. It was not. Matt reminded me that the virus they were hoping to use for ALS was very different from the one Wilson and his team used in Gelsinger's trial, and that it was widely considered to be safe, although any protocol that required repeated injections into the spinal cord carried enormous risks.

The death was tragic, Matt said. But why was the coverage on all the front pages? The edge of medicine is dangerous by definition. Matt was outraged that his whole field was being blamed and damned for one death. He often thought of the time he had been planning to treat a Canavan baby with his gene therapy, and the baby's parents decided to try a bone-marrow transplant instead. The baby died. If a nine-month-old child had died in his gene therapy trial, it would have made a scandal. But a death in an experimental bone-marrow transplant made no headlines, because bone-marrow transplants were not bolts from the future. No one ever called for a ban on bone-marrow transplants. So why should one death kill gene therapy? It was very sad, but it was also the very first death in a gene therapy trial.

Jamie felt the same way. He told me that he was disgusted by the scandal. He saw it as a symptom of a sickly society, a world grown so conservative, so timid about the edge of medicine, that it could hardly take a step. "From the first day you could just see everything that was going to happen," he said. "How people would take sides. Day one, you were going to hear the line 'The responsible members of the field say we should slow down to ensure this does not damage the field of gene therapy.' It was entirely predictable who would stand up and pontificate. And who would get killed."

Stephen and Wendy waited in their fourth-floor walk-up. Their political problems were local. In her memoirs of life with Stephen, Wendy writes a little more about their downstairs neighbor.

Not long after Stephen cut pads for the rugs, she says, he tripped on the edge of one of them and crashed into the bookcase. Within minutes, their neighbor downstairs was rapping at their door. Stephen had always been polite to her. But this time, he hurried to the door and flung it open.

"I am hearing an awful lot of noise from up here," the woman said.

Wendy, from behind Stephen, watched their neighbor peer into their apartment to see if they really had bought the rugs.

"I have a lot of work to do," the woman said, "and I wonder if you could keep it down."

"I'll try to do my best," said Stephen. And then, Wendy writes, she felt her calm young husband begin to break. She wanted to walk up and put her arms around him. She could see it all in his tall, broad back, as clearly as if she could see his face. All his agitation at his falls, and the crisis in his brother's race, and the knowledge that they would soon have to give up their dream of living alone together in Back Bay, the knowledge that he was bringing a child into a world that he might soon be leaving—she felt Stephen feel it all, and she felt him begin to lose it.

His voice climbed an octave and strained and cracked the way it did when he got upset. He told their neighbor that he had Lou Gehrig's disease, and that it was hard for him to walk because his legs were weak. The rugs made it even harder for him to maneuver, Stephen explained, because he could not always control his feet to lift them up. He kept tripping over the edges.

"I didn't know that," their neighbor said.

"No, of course not," Stephen said. "How could you?"

"God knows you never took the time to ask," Wendy hissed from behind Stephen.

"Well—" the woman said. Wendy thought she might have looked uncomfortable for just a moment. "Try to keep it down."

One week later, Wendy writes, their good neighbor downstairs was having some work done on the roof of her bay window. "Sparks from the roofers' blowtorches flew onto the roof and smoldered all day in the hot sun. By late afternoon they had turned into a blazing fire and ravaged her apartment. She had to move out for renovations."

That fire was savagely satisfying to Wendy, as if God himself were looking down on Beacon Street; but the word from Matt and Jamie was delay, delay. Jamie's race toward a gene therapy had begun exactly one year before, in January of 1999. He had accelerated for twelve months and now he was slowing down as he approached a wall of flames. He wanted to carry out the gene therapy with Washington's approval, but the whole brave new world of regenerative medicine was engulfed in political fire—set off by a scandal in gene therapy itself, in the city of brotherly love.

Jamie and Matt decided to put together an advisory panel of scientists to critique every aspect of their protocol so that each step would be as safe as possible. "I think the whole field is going to be scrutinized more carefully," Matt said. "We'll have to set up new procedures and be more conservative in terms of the amount of safety data we collect before each trial."

"I'm afraid we're going to have to put on the brakes a bit," Jamie said. "I can't tell you how much it kills me to say that."

{ THIRTY-FIVE }

Publicity

One morning at Rockefeller in the middle of the political firestorm in Washington, I got a fax from Charles Michener, my editor at *The New Yorker.* He wanted me to call him right away. Charles told me that because the decision of the RAC was pending, David Remnick, the editor-in-chief, wanted to run my story now.

I had planned to follow the Heywoods at least a little while longer—I hated to end the story with so much up in the air. But I went home and tried to get it done in ten days.

While I worked, *The New Yorker* sent a young German photographer named Martin Schoeller to Newtonville, and he spent a day with Jamie and Stephen on the second floor of the carriage house. In the shot that the editors eventually chose, Stephen sits in a ruined armchair in a black leather jacket with his chin up, looking proudly and levelly into the camera, like a king on a throne. Jamie leans over him, resting his folded arms against the back of the chair. He wears an elegant dark jacket and tie, and just a hint of his Supermanic half-smile. He might be the king's right-hand man. The space around them is nothing but darkness, rubble, rot, rusting bedsteads, and cobwebs.

In the picture, the Heywood brothers both look so handsome, young, and healthy that without knowing them it would be hard to guess which one could beat the other in a bout of arm wrestling. The story was a scramble for the editors at the magazine as well as for me, and on the Friday evening before they went to press, Michener and I were still on the phone going over our last changes. At seven o'clock or so, he called me back with his final question: "Just making sure. Is Jamie the one on the left, and Stephen the one on the right?"

In Boston on Monday morning, Jamie drove from newsstand to newsstand, kiosk to kiosk. He wanted the newsstand edition because the editors had announced the story in huge type across the top of the cover flap: "Curing the Incurable: A Brilliant Engineer's Race to Save His Brother's Life." Jamie bought all the copies he could find, more than three hundred. He was even prouder of the story than I was, and he knew how to use it.

John and Peggy loved the story, and the portrait of the boys seated in the ruins, although John did wonder about the symbolism. John also wondered what the words on the cover flap would do to Jamie's ego. He spoke to their family visionary in the dry, cutting, medicinal voice that everyone in the family had always administered to Jamie as needed. *You know, Jamie, these journalists may* say *you are a brilliant engineer . . .*

On the phone, Jamie told me that he and his small crew were distracted by all the attention. They were getting hundreds of crazy calls from quacks, zillions of crank e-mails. I quoted the old sign from down in the Mill Street basement, "Focus Focus Focus." Jamie said, "We have a new sign up: 'Fuck the Holistic Vitamin Vultures.' "

Someone on the RAC committee called me to say, off the record, how touching the Heywoods' predicament was, and how awful it felt to read about it just then, while sitting on that committee. The voice at the other end of the line sounded guilty, conspiratorial, and distraught. "They're all human beings," Jamie said when I told him about that call. "There are no evil people."

In Tennessee, the mouse was still alive. It was now one full year old. Jamie kept checking in with Marge Sutherland to ask after it. The Heywoods called it Minnie.

Jamie's publicity accelerated just as he had planned. A television producer, Robert Wallace, decided to do a segment about the Heywoods for *60 Minutes II.* A friend of the Heywood family, Richard Saltus, a reporter at the *Boston Globe,* finished a long feature for the paper, and the *Globe* ran it on the front page: "Hunt for Cure Gets Personal." A few Hollywood producers were interested. They called Jamie, his brother Ben, and me. But they were wary. Where was this story going? How would it end?

A pair of documentary filmmakers in Cambridge, Jeanne Jordan and Steve Ascher, asked if they could make a film about the Heywoods. They had just finished a film, *Troublesome Creek,* about the struggles of Jeanne's family farm in the Midwest. While they were editing it, Jeanne's mother's speech began to slur. They brought her to Boston to see Bob Brown, and he told them she had ALS.

Jeanne Jordan said that just after that appointment, she stepped into an elevator and found herself alone with Doctor Brown. *Is there any chance,* she asked him, *that it isn't ALS?* He did not answer her right away, and when she looked at his face she saw that he had tears in his eyes. Jeanne's mother died before they finished *Troublesome Creek.*

Now that my article was done, I was busy starting my seminar at Rockefeller, and dealing with crises in Providence. For my mother it was suddenly too late to hope for anything anymore, even from the edge of medicine. Things were going so badly with her that I had a few lunches that month with the philosopher Peter Singer at Princeton, to talk about the ethics of feeding tubes, living wills, and euthanasia.

Suddenly I was done with the Heywood story, which felt very strange. I kept in touch with them, and I followed the gene therapy debates. Some of the editorials that came out sound sensible and coolheaded to me when I reread them now. At the time, I took them as personal rebukes. The *Wall Street Journal* published a guest editorial about gene therapy. It was written by one of the *New Yorker*'s staff writers, Jerome Groopman of Harvard Medical School. When I read it I had to wonder if he was writing in response to my article. On the one hand I agreed with him; on the other hand I still hoped Jamie would save Stephen. "Expect progress, not miracles," Groopman wrote:

> Athena does not spring full-born from the head of Zeus. Rather, dreams become reality through hard labor at the laboratory bench and at the patient's bedside. Promises are fulfilled by cobbling together increments of knowledge drawn from the many corners of biology and medicine. A roller-coaster ride is dramatic and dizzying, but ultimately returns us to our starting point. Genuine scientific progress is made on a steady chugging train that at times feels frustratingly slow, but delivers us to our destination.

Bioethicists explored the balance in the Heywood story between hope and false hope. Eric J. Cassell, a doctor on the National Bioethics Advisory Commission, told one medical reporter, Vida Foubister, in an online weekly called *American Medical News* that "the need for hope is coercive." Some bioethicists suggested that "the most vulnerable patients shouldn't be allowed in trials, because they can't make a rational decision." Their arguments outraged Jamie. "That's just like totalitarianism," he told the reporter. "It's completely un-everything our society is about."

"I think that's one of the most condescending things I've ever heard in medicine," Jamie told me. "I mean, it's almost eugenic. If we've gotten to the point where human beings aren't capable of making decisions in their own best interests, then what kind of a society do we believe in? Don't you love a medical system that can produce doctors like that? I want to be treated by someone who believes that my opinion has merit."

Stephen issued a formal statement: "I believe that the last few years have seen a profound change in the way ALS is viewed by patients and doctors. I also believe that patients are living longer through a better understanding of the disease. The key to this is hope."

With Jamie, Stephen had a new line. *Now would be a good time,* he said. *If you're gonna find something, now would be a good time.*

If Jamie and his group did get permission in Washington for their gene therapy project, the surgeon on the team at Jefferson was still ready and willing to inject the EAAT2 gene into Stephen's neck. That is where the motor nerves that control the arms are situated, and those are the nerves that first started to die. Although Jamie hoped that the procedure would do some good for his brother, Rothstein had asked me to emphasize in *The New Yorker* that it was only a safety test, a highly preliminary trial in which the doctors would find out whether they were causing any harm. In the long chain of trials that Katherine Evans had diagrammed for me on her yellow legal pad, the chain that can take a big drug company ten years and a hundred million dollars to complete, this was only the first step.

Rothstein was still confident that he could enlist twenty-five ALS volunteers from his clinic at Johns Hopkins. After the procedure, he would give Stephen and the other volunteers a battery of tests to see if the genes had any effect. Rothstein would test Stephen by repeated isometric stress tests, checking the strength of his arms, his legs, and his breathing. The arm is the most reliable test. The key test would involve a pulley system that measured the biceps, the triceps, the finger flexors, and the wrist flexors. Rothstein would be monitoring Stephen's decline or improvement essentially by arm wrestling.

That month, Stephen was happy to have other things to think about. He and Wendy were getting married in Grace Church. Melinda and Jamie were expecting their baby soon, and by now Wendy knew she was pregnant. "We would have had children anyway," she said, after she and Stephen had their first prenatal consultation with the obstetrician. "But now we have to try to live out fifty years in the next two."

Stephen and Wendy got married on February 19. Peggy had said, *Nobody will come.* And everybody came. But she was right about one thing. Snow fell.

In a heavy New England snowstorm, Ben flew in. So did Peggy's sister, who had had multiple sclerosis for fifty years, since she was twenty-two years old. She was now seventy-two and in an electric wheelchair, but basically in good health. At the house on Mill Street, Ben, the invisible brother, was interviewed at last by *60 Minutes II.*

Stephen's architect friend and mentor in San Francisco, Mark Little, flew in from California with his wife and daughter. A colleague of John's flew in from Cambridge, England, overnight, and flew out the next day. Paola Leone was there. Duncan Moss, Jamie's best friend, the one who had never learned the rules of arm wrestling and wrestling championships, came with his wife, Nena.

Jamie was best man. Stephen walked joltingly but he could walk down the aisle without help.

When the time came for the reciting of vows, Stephen began in a strong voice: "In the Name of God, I, Stephen, take you, Wendy, to be my wife. To have and to hold—" For a moment, the whole story seemed to pass across his face. He choked, and glanced up at the church ceiling to gather himself. "—From this day forward," he went on, with a little smile of apology for the tears. "For better, for worse, for richer, for poorer, in sickness and in health. To love and to cherish, until we are parted by death. This is my solemn vow."

Wendy got through her vows with some trouble, too. The minister set the ring in Stephen's right palm. Stephen held it out to Wendy, and she took it and put it on her finger.

Everyone cried, even Peggy. ("My boys are good bawlers," she says. "I'm not. Only time I bawl's when I'm mad.")

Melinda read a sensual Anne Sexton poem, and John read a passage from the Song of Solomon.

At the reception, the cousins sang incredibly off-tune. They chose the same song they had sung at the beach when Wendy and Stephen got engaged. "Why do birds suddenly appear—"

Then a band played, with a priest on sax. And everybody danced at the wedding of Stephen and Wendy.

{ THIRTY-SIX }

Into the Flood

By now Jamie could not walk into his favorite supermarket, Bread and Circus, without answering greetings at the door and questions in every aisle. On some days he enjoyed the attention and on others he sent Melinda inside for him while he sank down low in the car, hiding in the parking lot.

"But it's nice I'm not just nobody," Jamie told me. "I'm no longer cold-calling the way I was back in the basement. Everybody knows who I am."

With all the publicity, he now found himself negotiating grants for one million dollars. He moved the foundation a second time: out of the ground floor of the Victorian and into office space in an old renovated mill in Newton Highlands. Having money and an office and a staff again made him strut. "We're really out of the basement now!" he exulted on the phone. "It feels really strong. Sandblasted walls. Exposed brick. Glass doors. It is quite lovely. I have a big office. Too big, everyone here says, bigger than I need or deserve. But I tell them, hey, you're all making more money here than you were before. . . ."

That was a note I would hear again and again as the ALS Therapy Development Foundation took in millions. Back in La Jolla, Jamie had said good-bye to the fabulous houses and the fabulous pools: *This is not what life is all about.* Now the choices ate at him. In the parking lot, he discussed the features of the latest Porsche with one of the trust-fund babies who had volunteered to work for the foundation. Then he strolled from the new Porsche to his own set of wheels that day, Stephen's old black pickup truck. He still wanted it all, and his longing for luxury led to more fights with Melinda. She wanted to sell the Victorian and the carriage house and move into someplace small, cozy, and quiet, a place where they could afford the mortgage themselves, where she could find the tranquillity to write, and to cope. She loathed the Victorian now. She thought Jamie clung to the house as if it meant something mystical to him, as if owning a fine house and a few fine things meant that all was not lost. Melinda wrote in her journal: "Would I be so frugal if Jamie did not buy the Armani suit?"

ALS-TDF was now a power among the many competing ALS family foundations. In my article, I had decided to say nothing about my doubts and worries. They were too vague—in the end, what did they amount to, as long as he kept on doing the right thing? My adventure in medical writing had helped Jamie get started, and now I could only hope that he would do something good. "Medical writers and editors for the general press don't intend to inflict cruelty on suffering people," the science journalist Daniel S. Greenberg has observed. "But that's what they often deliver." What would the Heywoods' story deliver? Everything depended on what Jamie did now—and he made me just as nervous sometimes as he had in the beginning.

A mixture of truth, fantasy, real promise, and total hype still intertwined throughout his conversations. It was fascinating to experience and infinitely renewable, and it was part of Jamie's core. I was alarmed and in some ways comforted to discover that other people often felt as queasy as I did. At scientific meetings, where Jamie hawked his foundation's program, more than one expert thought of the Talented Mr. Ripley, a young man able to assume a role and play the part to the hilt. Some thought of Professor Harold Hill, the Music Man. Your skin began to crawl a bit, listening to him, and scientists who knew something about his great subject could see how he embellished and faked his way through it. Sometimes you could watch judgments play across the faces of the neuroscientists as they listened to him talk the talk and walk the walk. The scientists thought: *He no doubt has himself convinced as well.* It was almost as if, just out of sight, something very fine was struggling there with something bad.

For his board of directors, he had chosen a high-powered and experienced group, but it was a group that would give him great freedom. It included John, Jamie's father; and George Hughes, a Boston lawyer who sat on the boards of a number of public companies. Hughes had volunteered his time from very early days to help Jamie set up ALS-TDF. David Searls, Jamie's cousin, was on the board; and Steve Fowler, one of Jamie's new friends from the world of ALS, the man who had tried the stem cell therapy.

Melinda went into labor on April 7, 2000, and Jamie was delirious at the hospital. And when their baby girl, Zoe, was back home at the Victorian, he could not look at her without thinking in terms of the spine and the brain. The proud father was also a genetic engineer and an alum of the Neurosciences Institute. "You can watch the intensity of her eyes, every day, as she nurses," he said. "You can *see* her brain wiring." Melinda snapped pictures of Jamie diapering Zoe with the phone to his ear. The birth of Zoe did not slow him down. He was not sleeping anyway. "For me sleep's more of an emotional issue," he told me. "I don't need that much. If I get too much sleep, I get depressed."

His gene therapy project was moving very slowly. Not only was the field superheated, but Matt and his team were having technical problems with the viruses, which were now being engineered by his lab in New Zealand. "Slow-me-down, annoying things," Jamie said. They still could not get the EAAT2 gene and its promoters to fit into the virus. The package of DNA the Auckland team engineered, following Jamie and Matt's specifications, was too big for the virus capsule. Jamie and Matt designed a smaller DNA package, but when they tried it on nerve cells in petri dishes, the cells did not make enough

EAAT2. Again and again the tests failed: If the engineers used the package they wanted, they could not get it into the nerve cells; if they used a package they could get into the cells, the cells would not do anything with it.

There were more rounds of meetings in Matt During's office, with the calendar pictures flipping now from month to month. There were many meetings at Jeff Rothstein's office in Baltimore, too, looking for clues in his latest unpublished studies of the EAAT2 gene, trying to learn from the natural engineering of the gene a clever way to reengineer it in the laboratory. The team tried a series of different promoters, with frustrating results. Jamie predicted that his EAAT2 gene therapy would be ready not in June but in September of 2000. "I hope that won't slip again."

Matt and Jamie met in Matt's office at Jefferson that spring and took stock. The neurovaccine was ready if Jamie wanted to use it. Matt's paper had been published in *Science* at last. In it, he and Paola and their colleagues showed evidence that a neurovaccine seemed to have protected rats from stroke damage. Matt told Jamie that he was ready to try the neurovaccine on ALS. But Jamie was still scared of Matt's neurovaccine. "Back at Advent, we had a rule," he said. "Only invent one new technique per project. The neurovaccine has a lot of inventions in it."

So Matt and Jamie decided to try the stem cells while they prepared the gene therapy. Their first test of the stem cell therapy, back in December, had not done any good for their dying ALS volunteer. Steve Fowler thought his speech had improved a little. If so, the improvement was not measurable, and his limbs and his breathing continued to get worse. But at least the stem cells did not seem to have done him any harm.

Doctor Brown was nervous about it, but Stephen told Jamie that he was willing to take the chance. In his apartment building, he took the stairs very slowly now. He planted his left foot on the next step up, hoisted his body up, swung his right foot up. Wendy went up behind him, but Stephen was six foot, three inches tall and that spring he weighed one hundred ninety-five pounds. The stairs were not easy for her either, she writes in her memoirs of life with Stephen. She was showing now: "Every day, I had more trouble hauling a loaded belly (which was pressing on my lungs) up four flights of stairs."

As they were going out one day, Stephen stepped off the third-floor landing with his left foot and misjudged the distance. Wendy had stopped for him on the second-floor landing, and she watched Stephen pitch forward full length toward her in slow motion. She screamed, and he thudded and rolled down the stairs straight at her.

"After all my imaginary training, all my mental images of catching him," Wendy writes, "I dodged him."

Stephen fell into the banister, flattened against it for an instant, and bounced off into the wall. He came to rest on the landing with his feet up on the bottom step and his back wedged into the corner.

Oh my God! Wendy screamed. *Honey, are you all right?*

His fall and her screams had raised an enormous echoing racket in the stairway. As she helped him stand up, she was glad it was a weekday. There was nobody in the building to help them, but at least none of their neighbors would see them: a woman pregnant by a dying man who could no longer even manage their steps.

What were we thinking?

Stephen and Wendy left Back Bay and moved back into the house on Mill Street; and Jamie decided to give Stephen the stem cells. It was now April 2000, and the gene therapy still looked far away. Jefferson's IRB ruled that because the procedure was being done to treat an individual patient, not as a research protocol, Matt, Paola, and Jamie were free to proceed. The procedure would be carried out by Fred Simeone, the chairman of Jefferson's department of neurosurgery—the same neurosurgeon who was prepared to carry out Jamie's gene therapy. Because the procedure was novel, all of the doctors involved were taking a certain risk.

Jamie sat down with his family in the house on Mill Street and they all talked it over, evening after evening. Stephen said he was sure he wanted to do it. Then Jamie sat down at Jefferson and talked with Simeone. He said, *This would be an amazing thing to do for my family. We don't know if it will work. You need to understand, you can look me in the eye. If Stephen dies on the table you will never get anything but gratitude from us for the rest of your life.*

They would have to put 25 million cells in Stephen's brain to get enough stem cells in there. If they waited until they developed the KDR technique, they would have to put in only three hundred thousand cells. It would be safer and more elegant, and it would be a merchandisable technique.

The way Jamie tells it, he decided he had to withdraw from CNScience. "I pulled over one day and said—I'm out. If I had proceeded to develop KDR, the therapy would have been delayed by eighteen months. So Matt and I closed down the company. I didn't want profit questions to come between Stephen's treatment and the right decision. I thought there were one-in-ten odds it would do anything. Which is why it was important not to do it as part of a company, for profit. You couldn't do it in corporate architecture.

"So we had a choice—wait and do KDR? Or go for it now?" The massive but inelegant injection might help. "No reason not to just try it—unless you were intellectually or financially interested in the KDR application—which is why I quit. When I went into a room as president of CNScience everyone wanted an angle." What's in it for me? "When I went into the room as founder of TDF, everyone wanted to help."

So in that way he went into the project with clean hands. But I think Jamie felt almost as worried about stem cells as he did about the neurovaccine. He would have asked Stephen to risk his neck for the gene therapy, but he was nervous asking him to do it for stem cells. Rothstein had some idea how EAAT2 worked; no one knew how stem cells worked. "It's not a cure," Jamie worried to me. "I don't know how effective it will be. It's lightweight."

The truth was that stem cells were still more a hope and a promise than a treatment or a cure. Art Caplan was just as discouraging about stem cells as he was about gene therapy. "My skepticism about all this is somewhere between unbounded and enormous," he told me. "A stem cell effort? Good luck! I mean, stem cells, I'd put that on the five-year plan, not the short-term plan." Jamie was more likely to hurt Stephen than help him. "You can actually kill somebody," Caplan told me. "You can actually make them suffer more miserably. Somebody might say, 'How could you suffer more miserably than by having ALS?' But you can. Way back when the artificial heart was being developed, in the mid-'80s, I watched Barney Clark, who was dying of heart failure, die instead of nosebleeds, dementia, psychoses. He just had a miserable dying because the device wasn't ready. They rushed it to save him, and he agreed to try it because he was doomed. He certainly was going to die—faster than this gentleman is. But the fact is that you can be made more miserable—even though it doesn't seem possible—by interventions."

As the date for Stephen's injections got closer, Jamie sounded more frightened. They had given human stem cell injections to two monkeys without doing any obvious harm, but that was almost all the data they had. "This is risky," Jamie said. "This could be worse than Gelsinger. It's really borderline. We could cook him. Jeff Rothstein thinks I'm crazy, outside the curve."

That was true. "Wow, are you kidding?" Jeff Rothstein said when I asked him what he thought of Jamie's plan. "Putting foreign cells into the cerebrospinal fluid—cells that are not normally there? They could cause an inflammatory response and become horribly toxic. Jamie keeps telling me that this is just a variation on an FDA-approved procedure. Yes, it is, but you put the patients' stem cells back into their blood. You don't put them into their spinal fluid. Years ago, some doctors injected formaldehyde in there by accident and killed someone. You don't put things in there. That's a sacred space in your body. What Jamie is planning has never been done in a human being before. It's just ludicrous. I'm appalled. Jamie's decided he is going to practice medicine on his own for the sake of his brother."

Rothstein said he hated to sound like a stereotype—like the monstrously overcareful, uncaring doctors you see in the movies. "Of course we care," he said. "We just have a different view. We recognize that people go into these experiments extremely biased, thinking that no harm could come of them. They say, 'What's the harm, I'm going to die anyway!' But there can be harm."

On Sunday, May 21, 2000, Stephen and Wendy stopped by the Victorian before they left for Philadelphia. They were driving down with Peggy—Jamie would follow after he took care of some business at the foundation. Stephen gave Jamie and Melinda long hugs. He said, trying to joke, *Let's see if I come back dancing and jumping around.*

On Monday, Wendy and Peggy drove Stephen to Jefferson Medical Center to give blood. Normally the doctors would have taken it from his arm, but Stephen felt so weak that if they stuck him there, he was half afraid he would never be able to flex his arms again. Besides, he did not think he could make a fist. The doctors decided to insert a catheter into his jugular instead. They also gave him a drug that flushed the stem cells out of his bone marrow.

Stephen, Wendy, and Peggy stayed that night at Paola's three-story row house, with the cat, the rabbit, and Gnocci and Bambi, the German shepherds. Gnocci and Bambi greeted them at the door and were extremely excited to see them. When they put their paws on Wendy's shoulders, they were taller than she was. She felt protective of her belly. She was almost six months pregnant.

On Tuesday, she and Peggy took Stephen back to the medical center, where doctors drained blood from the catheter they had implanted in his jugular. Stephen watched movies on a VCR while his blood drained, Peggy wrote thank-you notes for the foundation, and Wendy read, read, read—she was trying to read as many books as possible before the baby arrived. In a separate laboratory the doctors extracted what they needed from Stephen's blood and then returned the rest through the catheter. Meanwhile, with the extract, they stripped away as much as they could to concentrate the stem cells. Simeone, the surgeon, had stopped by the row house and given Stephen a statement to sign:

> May 23rd, 2000
> Stephen Heywood's Statement
>
> I Stephen Heywood voluntarily write this statement to:
>
> Dr. N. Flomenberg
> Dr. A. Freese
> Dr. M. During
> Dr. P. Leone
> Dr. F. Simeone
>
> I was diagnosed with ALS in December 1998 and the disease has progressed rapidly in my body. I have been following the research being done in different laboratories on Stem Cell transplants. . . . I request that you transplant my own [stem] cells into the intrathecal space of my brain and my lumbar spine.
>
> Because what I ask is a novel procedure I take all risks for any possible outcome and I will be responsible for any personal and/or institutional liability for the result of the procedure. I am personally asking for this procedure and do so under no pressure or duress and I personally take all responsibility for the consequences.
>
> I also understand that any novel procedure involves risks and that in this case though there is no known reason to believe there will be negative consequences it is possible that unanticipated consequences could result in partial paralysis or even death. I am conversant in the technology and operations involved. I understand the

> potential risk is real and I choose to accept it as reasonable. I also understand that side effects of the operation can include spinal infections, headache and a potential blockage of the spinal fluid. I choose to accept these risks.
>
> With this statement I also fully agree to allow post mortem analysis on my brain and spinal cord in case death would occur as a consequence of this procedure.

At the bottom of the page, Stephen Heywood scrawled his initials with his left hand. Jamie Heywood witnessed the signature.

I talked to Jamie on the phone that night. I had never heard him so nervous. "We have no clue it will do any good," he said. "And there's some risk."

In his fear, Jamie sounded almost as skeptical of his own project as Art Caplan. "Everyone wants one thing that will do it—like *acupuncture,*" he told me. He pronounced the word with the same contempt that Caplan had used when he compared gene therapy to entrails and bat wings. That night Jamie felt terrified of stem cells. He was asking Stephen to risk his life for something that would probably do him no more good than a placebo.

"But," Jamie said, at last, "if he *believes* it's powerful . . ."

On Wednesday, May 24, Stephen was taken into the fluoroscopy room at the medical center for the procedure, which would take two hours. Paola and Matt assured Wendy and Peggy that the neurosurgeon who would inject the stem cells into Stephen, Simeone, had performed more spinal operations than anyone in the world. They would not let Wendy stay with Stephen, and she hovered outside the door of the fluoroscopy room. At one point the door opened and she saw Stephen lying on his side on a gurney with his back to her. She saw Jamie, too. Jamie, who always had a way of getting where he wanted to go, was hovering above Stephen with a video camera, taping the procedure. The room was tinted green from the fluorescent lights. The door closed again, and for the first time Wendy felt a thrill of horror instead of hope.

Simeone, the surgeon, injected 20 million stem cells into Stephen's spinal fluid. Then, using a spinal needle that he guided with the fluoroscope, he injected more stem cells into the cisterna magna, a big space in the back of the brain where a surgeon can insert a needle with relative safety. It is at the back of the lower part of the skull, right where the skull and spine meet. Stephen felt fine when they were done, but he had to stay at the hospital overnight, and the hospital rules would not allow Wendy to stay with him. She went back to the row house, said good night, lay in the dark and could not close her eyes. She wanted Stephen spooning her in the bed, telling her that he loved her, and telling her that they were going to be fine.

On Thursday, May 25, Wendy and Peggy picked up Stephen from the hospital, brought him to Paola's, and set him up in an upstairs room, away from the dogs. Then Wendy and Peggy walked down South Street, where they bought a thank-you gift for Paola, a tall white orchid. Carrying the orchid back to Paola's, Wendy stopped in front of a store. In the window she saw a huge steel canopy bed. She was in such a state by now that the bed seemed like a sign. She and Stephen wanted a bed just like it. They had been shopping in Boston for just that bed. Stephen had even begun to think he would have to make it himself. The bed in the window had handsome two-inch-thick industrial steel tubing, and it was high enough that her husband with his long legs could get in and out of it easily. Even the price was right: only $600. Wendy is not a matter-of-fact atheist like Stephen. She is a doubter, an agnostic who often argues and pleads with God, and back at Paola's that night as they all drank wine, she talked about the bed in the window almost as if it might signify a miracle.

Matt and Jamie played pool. Stephen watched and teased. But Paola found the celebration almost too poignant to bear. She and Matt had no reason to think the stem cells would do more for Stephen than they had for their first volunteer back in December. This was just one of the first incremental steps the world would have to take in developing stem cells into useful treatments. The amount they had to figure out was more than the amount they knew, and so much could go wrong. And even so, Paola hoped and believed. It was quite typical of her to have hosted the Heywoods in her home, with her dogs, cat, rabbit, with more stuffed animals and stick-figure drawings like the ones she kept at the lab—gifts from toddlers with a year to live. Paola puts no walls between herself and her patients, as Matt does, and she suffers like one of the family.

On Friday morning, when Stephen, Wendy, and Peggy sat down in the kitchen for breakfast, they found an envelope on the table: "To Wendy and Stephen." Inside was a card from Paola and a check for six hundred dollars.

{ THIRTY-SEVEN }

The New Model T

John and Peggy Heywood helped the newlyweds buy a little Cape Cod three-bedroom house in Newton on Beverly Road, just a few minutes' walk from the Victorian and the carriage house. They were still paying for those, and they were bankrolling Stephen's work on the carriage house.

The whole family helped them move into the house. Stephen and Wendy had fun assembling their steel bed, and like any expectant parents they threw themselves into renovations and decorations. But Stephen noticed no improvement in his condition. Bob Brown examined him at Massachusetts General, and he saw no improvement either. Stephen's right hand and foot were still weak, and still getting weaker.

The baby was due in September, and Wendy could not help feeling frightened now. On most days she did not panic about the one big thing—she panicked about many little things. She was still working at a Harvard biological laboratory, and she came home fretful every evening. "Dear God," she scrawled in her journal, "Why are you killing Stephen?"

Stephen himself stayed calm. He kept his sardonic sense of humor, and he did not lose his perspective even now. The two of them were still there, and the baby was coming soon. Whenever Wendy said, a little overearnestly, *Stephen, there's something I want to talk about,* he said, mock-anxiously, *You're pregnant?*

Six years before, Wendy had started and abandoned work on a master's degree at Southwest Missouri State University in Springfield. It was an MA program in English, with an emphasis on writing. Stephen urged her to call the school and find out what she had to do to finish. With his encouragement, she proposed a project that might qualify as her thesis. And that is how she came to begin her memoirs of life with Stephen. Stephen's suggestion was inspired. Instead of fretting in the long summer evenings, Wendy could celebrate their story, and their family's, and she could haul a few of their neighbors in Back Bay before the court of literary justice.

One more ALS patient, a forty-seven-year-old woman, tried the stem cell injections. Once again the injections had no measurable benefits. All three patients went on getting worse, and none of them chose to repeat the procedure. But at least nobody seemed to have been hurt, and all three of them felt that they had helped the cause of regenerative medicine.

Afterward, Jamie explained the failure of the stem cell work to me this way. "If you think of stem cells as a repair crew, ALS is a flood region. Can't send in repairmen now. The flood is still there." He thought he and Matt and Paola needed to clean up the glutamate flood before they could put new nerve cells in there and see the cells grow and go to work. That would take his EAAT2 bailers.

According to another view of ALS, Jamie's explanation was overoptimistic. By the time motor neurons begin to die in ALS, the glutamate flood is probably gone. It has rolled on. It is elsewhere in the spine. The flood is gone, but the nerve cells that it damaged have begun what is known in the jargon as apoptosis, a process in which cells give up and die. The nerves in Stephen's spine had begun literally killing themselves.

Jamie was beginning to lose his faith in futuristic medicine. He still hoped that his gene therapy could correct Stephen's problem at or near the source, and stop the last healthy motor neurons from being damaged. Then, someday, in five or ten years, he could use stem cells to replace the nerves that had died. But in Washington, gene therapy was now absolutely untouchable. Paola got massacred when she presented her Canavan protocol at the next RAC meeting, Jamie told me. "People on RAC are *posturing!*"

Matt decided that they might do well to wait awhile longer before submitting their ALS gene therapy proposal to the RAC. Maybe the world would be calmer in September. If the RAC approved their plan then, Jamie calculated that Stephen might get the EAAT2 gene in the spring of 2001.

"It's like molasses," he said. "There are so goddamn many things in the way."

With Steve Gullans of Harvard Medical School, Jamie had begun to look in still another direction. Gullans was convinced that there might be buried treasure in the medicine chest: They might find cures for the incurable in drugs that were already on the market. Gullans was trying to develop and automate a way to test drug after drug on cells in petri dishes or on mice to see what they might do for orphan diseases. Jamie could go prospecting in the tailings of old gold mines while he fought to open up the new frontiers.

Jamie's foundation, a few minutes from the Victorian, had about three thousand square feet. On his office wall there was a framed black-and-white picture of Stephen and Jamie as boys, and a photo of his little baby Zoe on his desktop, behind stacks of files. There were also slogans written graffiti style on the glass wall of his office: "Time Is Neurons!" "Trust but Verify!"

He was already thinking of moving again. He needed an animal facility. On a table behind the reception desk stood stacks of plastic cages with rotating wheels. He had designed them himself. Each one had a built-in data collection unit, and a fan to clear out the ammonia from the mouse litter. A company was building the cages for them at cost. With individual data collection and micro-controllers, Jamie thought they could do automated checks of the energy and the strength of the mice as they tried out drugs on them in different combinations.

Jamie kept trying to get me excited about his new idea, but every time we talked I felt sadder.

"So you're no longer the guerrillas charging the hill," I said.

"We're moving away from the guerrilla metaphor. And from *Brave New World*." In fact, Jamie told me, he had begun to think about the Model T.

"Henry Ford had this insight: I don't need to build a Dusenberg. I just need to build a simple car that gets from A to B." What if the scientists and technicians in his foundation could automate tests of all the drugs ever approved by the FDA, and try them out one after another on ALS mice? If they found something that worked—or a cocktail of drugs that worked in combination—there would be no red tape, no delays, since all these drugs were already approved by the FDA.

Jamie tried to sell me on the Old World the way he had sold the New. "I remember when we were all going to the moon," he said. "Rockets! I was going! I was born in '66. Air flight has not changed since that day. Not at all. Military jets have not changed. We're not in rocket ships." The Brave New World in medicine was still coming, he was sure of that. It would come as soon as we were ready to make it come. So would the space colonies. "We *could* build a rocket ship to Jupiter. And medicine *will* get better. Gene therapy *will* be done—and it will be just another drug. Not Brave New World. Stem cells—we'll figure out how to do it. ALS, Parkinson's, Alzheimer's will be cured. Not by miracles. You'll go to the pharmacy."

Meanwhile he had found an open niche, he said: research that should be done but that most big biotechs would never try. "Because it won't make Genzyme any money to show that, God forbid, aspirin helps with EAAT2," he said. "They're interested in what makes money."

I tried to feel excited by Jamie's "back to the past" speech, but I only felt depressed. It was now late summer 2000. Almost a year had passed since I had climbed the ladder to the hayloft of the carriage house with Jamie and Stephen. I remembered how hopeful we had felt and the gallantly sardonic way that Stephen stood up there, with folded arms, and surveyed the ruin: the rotting floor, the smell of rats and rot, the spider-shrouded dormer windows.

"So this is a bit of a death trap," he said.

And Jamie laughed. "No, it's not," he said. "This is our house."

{ THIRTY-EIGHT }

A Sick King

Each time I visited Providence, my mother looked smaller and stiffer. She was slowly bending over into a question mark. To meet her eyes I had to sit at her feet and look up into her face. When we connected, she gave me a locked stare. In her checkup in the fall of 2000, Doctor Salloway held up a finger in front of her face and moved it to the left and right. Her eyes tracked it. Then he moved his finger up and down. Now she followed it by tipping her head, not moving her eyes. Placing his hand on her head to hold it still, he raised and lowered his finger again and again, and her eyes just stared straight ahead. When he tipped her head forward and backward, very slowly, her eyes rolled so that her gaze stayed dead level, like the eyes of a doll.

Salloway looked at my father. *I think I know what this is.*

First her rag-doll, forgotten-marionette falls, then the slow stiffening of her body (just getting her outside was a big production now). With that suite of problems, and now this peculiar selective paralysis of the gaze, the diagnosis was clearer. It was not Lewy body dementia after all. She had Steele-Richardson-Olszewski syndrome, also known as progressive supranuclear palsy, PSP.

Steele was the young doctor who had gone to the island of Guam to work with the native Chomorros and study their peculiar epidemic of nerve-death disease, which manifested sometimes like ALS, sometimes like PSP.

In PSP, the gunk in the sick neurons is not made of alpha-synuclein, like Lewy bodies. It is made of tangles of tau, one of the same proteins that accumulate in Alzheimer's, although it piles up in a different set of cells in the brain. So the vaccine that Matt and I had talked about would have been a long shot at the wrong target. But I did not go back to Matt to talk about tau. By now the experimental neurovaccines were looking almost as uncertain and unready as gene therapy; and if she had ever been a candidate for a trial at the edge, my mother was too far gone now.

Jamie wanted me to keep writing about his foundation. He called often.

"Come down to Duck with us. Everyone in my family read the article. They all love you."

But I was too sad and demoralized to follow his progress.

Once when we saw each other, I told Jamie that my mother's disease was probably PSP, but that we would not know until the autopsy. And Jamie laughed. "I'm sorry," he said. "It's just all so ghoulish sometimes."

My title at Rockefeller was writer-in-residence, but I was not really in residence. I still lived out in the country in Bucks County, Pennsylvania. Once a week or so, I went into the city on the bus. Actually, I was not writing, either. I was too depressed. After all our hopes, I saw nothing but bad signs. It was the onset of a time that I and some others remember as a long dark blur. The Internet bubble popped, and the biotech bubble popped, too. Matt and Paola, who had been seeing each other secretly, had a horrible public falling-out, and Matt had her escorted from the CNS Gene Therapy Center by security guards. Then, in the middle of a complicated squabble with its department of neurology, Jefferson closed down Matt's center, and he sued the school for breach of contract. Margaret Sutherland moved her lab from Vanderbilt to George Washington University, and on the way to Washington, Minnie died. Sutherland would not publish anything about the miracle mouse, so a student of Jeff Rothstein's tried to replicate her experiment. His ALS mice took a little longer to get sick, but they did not live one extra day.

The fall of 2000 was happy for Stephen and Wendy, at least. Their son Alex was born by cesarean section on September 11, exactly one year before that date turned dark.

As before, I hung around in the laboratories of scientists, and followed their patient work and their more than patient view of time. Günter Blobel and I often talked that year. One afternoon Günter was reclining across two or three chairs at the foot of the conference table in his seminar room, relaxing as he chatted genially and comfortably, rubbing his eyes.

"If we knew the signals, we could replace your heart tissue," he said. "Say you had a heart attack. I could take stem cells from you, take out the nucleus, put it in one of your heart cells, put it back in your heart. Then all the proteins the cell makes will be *your* proteins. Then you can repair the damage in your own heart! This has been done with mice, it was published in *Nature.*

"This will be a very important line of research in the next ten or twenty years. I only regret that I don't have another thirty years of creative work. I would do another long-range inquiry like in the past. I am always interested in the long-range inquiry—not the bread-and-butter questions that give you answers and that's the end. But I know I don't have thirty years of creative work. You have to just face reality and hope for the best. To do science, you must never lose a child's hope," he said, a line that became a proverb for me. "You must not fall into the cynicism of old age, that nothing matters. Once you have lost that childish hope, then you have lost a very important element for doing science."

A long time before, when he was just starting out, Günter had asked a scientist for a letter of recommendation, and the man had written that he was immature. Günter found out about it, and now it amused him vastly to remember that letter. "'Clever, smart, but immature!' Of course, I didn't get the job! He mistook my childish enthusiasm—he felt this was immaturity. Probably he had long since lost hope himself. Many do. Psychosclerosis, you can call it. Some have it at age twenty. Many would consider them mature. Probably they are! They have nowhere to go! Other than falling off the tree and being eaten by the worms."

He leaned his head back, hands clasped behind his head, on the long row of chairs. He wanted me to go to Dresden, he said. I had to see the way the Dresdeners were restoring their ruined city and I had to stand inside the Frauenkirche. "You have never seen a space more magnificent than that space! Reconstruction has spiritual meaning. Buildings are not just brick and mortar. It doesn't matter if the brick and mortar are five hundred or three hundred years old, or new, what's important is the *spirit.*

"Einstein said, 'Religion without Science is blind. Science without Religion is lame.' The Communists tried to do without religion and it was a catastrophe. We've seen what the Russian rejection has brought their cities and their citizens—total devastation.

"You can make your own religion if you want—and you can look at religions as a struggle to deal with life through metaphor, to deal with what we cannot explain. The great old solutions are worth studying, and their similarities are worth exploring.

"The cell arose three and a half billion years ago! So as we sit here we are really three-point-five billion years old! So for practical purposes we can really speak of eternal life! And the Resurrection!"

I did not see the Heywoods again until May of 2001. That month, the story of Jacques Cohen's experiment, the project that I had heard about in the Caribbean three springs before, broke at last. It was one of Cohen's own people at St. Barnabas, the young scientist Jason Barritt, who forced the story to the world's attention. He was leaving the clinic at St. Barnabas, he was looking for a job, and he published an article announcing that they had been the first to genetically engineer human babies.

The press took up the story in the United States, in Europe, in Asia. The angriest articles were published in Scotland, the birthplace of Dolly. The *Scotsman* ran a story about Jacques Cohen on May 9, 2001, with the headline "GM Baby Doctor: My Little IVF Experiments." Next the *Scotsman* ran an editorial, "Birth of a Crime Against Humanity." In the twentieth century, the editorial said, political utopias had led to murders. In the twenty-first, would scientific utopias lead to murders?

> As the criteria for perfection become ever more narrow, couples who produce ugly children of moderate intelligence through opting for the "outdated" method of conception involving love and luck, will find themselves charged with committing crimes against humanity. Ways will be found to deal with such aberrations.
>
> "It will never happen," I hear you say, but that's exactly what we used to believe about GM babies.

Cohen lives on the Upper West Side of Manhattan, and one morning he stopped by my office at Rockefeller on his way to the clinic at St. Barnabas. He sat on the edge of the sofa, brought out a silver pillbox, and swallowed a few pills. He sounded out of breath, as if he had been running. Not only were the world's tabloids accusing him of genetically engineering the Brave New World. The *Washington Post* was now about to announce that several of the babies conceived with the help of cytoplasmic transfer had genetic problems.

"The critics say our kind of work should be risk-free." He pointed at his silver pillbox. "I'm taking a drug for acid reflux disease. It's a wonderful new drug, especially for the last two weeks a godsend. Comes with a list of risks. I am the experiment, no? I am the experiment. And who knows what it does to me.

"I don't think we did anything wrong. I think we did the right thing. This is just bad journalism, irresponsible reporting. PR at St. Barnabas has been working on this seven days a week. And the legal department, too. The reactions of all the journalists! Some of them are claiming that this is the same thing as cloning. In Taiwan that's what they said. It goes around the world in twelve hours. They only talk about cloning. I'm 'Mister Cloning.' It's almost a joke. What we do is separate from cloning. But psychologically it seems the same to people. It's all needles and eggs. And scientists are all mad and irresponsible," he said, with a sudden tightening of his voice, as if his chest were caught in a vise. "They throw it all in the same heap. And of course the majority is *fertile.* So the infertile have very little say."

"You have my sympathy."

Cohen gave me a trapped, hunted-animal smile. He looked like a man who has been tortured in a stone cell, and then is told to get dressed and have a candlelit dinner with his torturer. There was a strained, battered look on his face and in his whole manner. Yes, you are being nice to me now, but are you going to hit me again?

Fortunately, Cohen said, even when a scandal is global, most people on the planet do not hear about it, even the parents of the babies in question. "I spoke to two mothers, they had absolutely no idea what had happened. Because they have two-year-olds—how do they know what happened! Mothers of two-year-olds do not watch the news! And it is nice to find that some of my colleagues have not heard of it. It is actually a pleasure."

"Do you still have patients coming in, this last couple of weeks?"

"The last couple of weeks? Hundreds, maybe thousands. Why, do you think the news scared them?"

"Yes."

"It might have scared me and you, but it didn't scare patients."

I walked Jacques out through the stone gate at 66th Street. As we were crossing York Avenue, I said, "Do you ever think about this, Jacques? Here you are making these decisions case by case, for the good of patients right in front of you. But when you see the babies that you have helped to make, do you ever think that somewhere down the road, this work, or even these very babies, could be the start of a whole new line of the human species? That this could be the beginning of a transformation?"

"No. It is too day-to-day. It's day-to-day. It's this hour, this patient, these clients, their wishes, this baby. I don't look past that."

"But you do sometimes take the long view," I said, thinking of the retreat on the *Galaxy.* "Don't you think it is somewhat eerie that you may be playing a role in some altered future?"

"No. Each one of us is such a small pea in the process. Every scientist makes such a small contribution. We are all part of a very large process. And so one doesn't think of oneself as playing any large role in that."

And the truth is that you can spin that project of his either way: as just another surgical procedure, or as the cosmic beginning of the transformation of the human race; as one more modest way for a doctor to help a patient, or as a new, revolutionary, and possibly fatal means for human beings to play God.

We were passing a big elementary school, PS 183, with a great stone lion above the arch of the doorway and wire mesh on the windows. We could hear children shouting and laughing inside. A computer-printed sign on the door hoped they would have a good summer. Abruptly, although we had not reached his car, he shook my hand and said goodbye. I had upset him. It was one thing to talk about the fate of the species on a Love Boat out in the Caribbean. Now he was in a global scandal. He was standing on 66th Street, watching his back. He looked terribly nervous.

"Well, it is a fascinating story," I said. "You will certainly get past this."

"You are promising me."

"I wish that I could."

"At the moment, it doesn't feel that I will get past it."

The next week I drove up to Providence, and from there, with dread, to Newton. Jamie was waiting for me in his office at the foundation with the sign on the glass: "Time Is Neurons!" He told me that they were investigating the claims of a doctor in Puerto Rico about gamma globulin. The doctor claimed to have tested it on six ALS patients. "If you believe the data, four out of the six have significant slowing of the disease. We've started a mouse study. The data from what's been done so far with humans is unclear. It's not even really a study—only four months or so."

He told me that he had gotten gamma globulin approved in eight hospitals in the United States for compassionate use. "If it works or not, I don't know. But we're managing clinical sites across the country, sharing our experience with everybody. This puts us in a weird quasi-regulatory role." Jeff Rothstein, in Baltimore, did not approve. "Jeff is mad at me," Jamie said. "He thinks I'm out of my mind, I'm going to hurt somebody. But what's the ethical response? At the moment, we're drawing the boundary where patients want us to draw it. Their decisions are their own. We're giving them information."

"Will Stephen try gamma globulin?"

"Maybe we should. The problem is, if Stephen does it, about two hundred others will."

"Why?"

"Well, everyone wants to know what he does. They call and ask."

Jamie and Stephen had become leaders in this adventure, just as they used to be in the laser tag games in the nights out in the woods when they were kids in Newtonville.

We drove over to Stephen's Cape Cod cottage.

"Stephen's a mess," Jamie said. "You'll be fairly shocked." They were still praying for some long shot to work. They needed a new road, he said. Going through the stem cell adventure had taken a lot out of all of them. "You can try three or four things, then you can't do any more. You become wary of investing hope. So you get conservative. And ALS hasn't gone slowly with Stephen. He's progressed much faster than I expected."

I knew what he meant about investing hope. I had invested and lost, and I felt it would be painful to hope again—but it was also painful to be there without hope. As we walked up to Stephen's door, I confessed to Jamie that I was nervous to see him again. "Do I try to hide my feelings?" I asked.

"I always reach out to take a patient's hand, even if he can't lift a hand himself. I ask, 'How are you?' "

"Okay."

"You can be honest. It pretty much sucks."

Stephen's small house had an addition in the works. There was a smell of plywood. I greeted Wendy and her new baby, Alex. It was a little while before I noticed a man in a wheelchair in the living room. He was wearing a baseball cap. The man was so thin and motionless and he was so low to the floor, sitting in the chair, that I had not spotted him. I did not want to stare, but even when I glanced once or twice at his face, and saw him staring at me, I was not sure it was Stephen.

I said hello. Of course it was Stephen, and of course I did find it hard to talk. Stephen said very little, too. He looked skeletal, and I think he was accustomed to letting old friends get used to the sight of him.

Jamie and Wendy showed me around the ground floor of their house in progress while Stephen wheeled after us. The wheelchair was motorized, with a control panel under Stephen's right hand, and he followed us on the tour of his new house. Stephen said something so softly that I did not know he was talking to me. Jamie translated. "Stephen asked you if you're getting the new titanium Powerbook."

Stephen looked at me and raised his eyebrows by way of a smile. Then he spoke, and Jamie spoke with him in simultaneous translation.

" 'So, Jon, this is the former garage. . . .' "

The bedroom was dominated by the steel bed that Paola's check had paid for. It was a fine bed. Stephen gave me another wry, charming look as Jamie showed me all the fancy equipment around it, one machine to help clear his phlegm, another machine to help him breathe. Stephen raised his eyebrows and widened his eyes a little, and grinned, though those muscles did not cooperate very well anymore. It was as if he were saying, *Wow, huh,* but with irony and good humor. He was still there.

Jamie, Stephen, and I headed out to a restaurant, Stephen leading the way in his wheelchair. He had found a shortcut through the woods. He had a rakish, Huck Finn style even now, I saw, even in this motorized wheelchair, hustling along through the woods.

"You need Weedwackers with that, Steve," Jamie said.

Steve said something and Jamie translated for me: "I don't really know where this ends up." The path had woods on the right side and a steep drop on the left.

"Do you worry about rolling off the left side?" Jamie asked.

Stephen ignored him, rolling at a good clip. We hurried behind him.

"How would I actually stop you if you did?" Jamie asked.

Stephen said something that sounded like, "If I could ride a Harley-Davidson, I can ride this." I was beginning to be able to understand him.

"So now I know if you disappear for a few hours where to find you," Jamie said.

We hustled along in silence through the woods.

"So I saw a lot of wheelchairs on Sunday, Stephen," Jamie said. He had given a lecture to a big audience of ALS patients.

"How'd it go?"

"Amazing in some ways. But I felt a little uncomfortable. As if I were a motivational speaker."

"Yeah."

"I don't know. It's almost like they cared more that I cared than that I was making progress," Jamie said. "I felt uncomfortable about it. I've got a new line in my office: 'Negative language, strong data.' Be very neutral in our language. Steve," he said, "you need to wash your chassis."

We came out of the woods and I saw that we were just around the corner from a restaurant, Naked Fish, a trendy upscale place in Newton. The place had a wheelchair ramp, and Stephen rolled up. Everyone in the place studiously and politely avoided looking at us as we progressed to our table.

Jamie talked about the gamma globulin doctor as he fed Stephen shrimp.

"What's he like?" Stephen asked.

"Good-looking. Smart. Very smart. He's saying some patients got better. My intuition is screaming at me to watch out. He has patients writing articles to the world saying, 'This works!' Most doctors would call that unethical."

Stephen got a disgusted look on his face. There is a sick king in Shakespeare's *All's Well That Ends Well* who has persecuted time with hope, and has lost hope with time, and has abandoned his physicians. Stephen had not abandoned his physicians, but he looked very tired.

{ THIRTY-NINE }

The Scream

In August of 2001 my father called to tell me that my mother was wailing one word over and over all day long: "yes."

I could hear her in the background as we talked, and I wondered how my father could stand it. But Dad was holding on. "Even now she really seems, much of the time, quite rational," he told me, while from the next chair in the sitting room I could hear her crying "yes, yes, yes."

Poor Dad tried to keep the conversation normal. Sometimes my mother could still get out a few words besides "yes," he said. And he told me the latest story about Grace. By this time, I had persuaded Dad to hire someone to help him with my mother. In Allentown, I had located a woman who had experience caring for the elderly and good references. Grace was willing to move to Providence if she could bring her sixteen-year-old daughter. So I had driven Grace and her daughter up to the house and helped them move in, and now my mother hated Grace.

My father told me that earlier that week the four of them had sat down to dinner, Grace and her daughter, my father and my mother, my mother repeating "yes, yes, yes."

My father said, "Grace, this is good chicken." And Mom stopped wailing for a moment. "He's just being polite," she said.

At the end of the meal, Grace smiled. "Well, Mrs. Ponnie, how did you like the chicken?"

Again my mother stopped wailing for a moment. "C plus," she said.

By the time I could take a train to Providence, a week later, the wails had turned into screams. The screams were even harder on my father, with his sense of decorum and his Old World horror of dementia. He hated to think that the neighbors must be hearing her through their windows. The weather was stifling that August, but he kept all the windows closed.

When I got to the house, Grace was feeding my mother Cheerios with a spoon at the kitchen table, and my mother was screaming. The whole house was filled with her rage and protest, a sort of animal protest that did not seem to comprehend much of the world—not this world.

The house was baking in the heat with the windows closed. We ate our lunch together, my father, my mother, Grace, and me. My mother's eyes were a dull gray, almost colorless. I wondered if even the color of her eyes could have changed. The whole house felt weird. As we began to eat, my mother wailed again. Then Grace happened to look down at the floor.

"Oh, I'm sorry, Ponnie," she said, with a laugh. "I didn't know that was your foot. My foot touched hers," she explained to my father and me. "I thought it was the table." That was why my mother was wailing just then, to tell us about the latest torment from Grace. We all chuckled, and my mother joined in with a kind of complicit giggle. She thought it was funny, too.

After lunch, Grace took my mother up to the sitting room on the second floor so that my father and I could talk downstairs in the living room. The wing chair in the living room belonged to my mother now. He sat down on the couch by the window. I took the love seat by the wall of books that I had helped to unpack from their boxes when we first moved into that grand old house: the shelves of Boswell and Johnson, the first-edition *Ulysses,* the poets, the rows of dictionaries and reference books that he had needed at one time or another: Russian, Italian, French, German, Latin, Yiddish, English.

Even from there, with the sitting-room door closed up on the second floor, I could hear that expression of longing, the yowl that began as the most conversational, casual, calm "yes," and then rose and rose, sometimes in anger, sometimes in rage, sometimes in grief or in a somewhat softer sadness. When a cat or a newborn baby wails like that, you wonder how they know how to hold your attention so well. They vary themselves just enough that they force you to follow every note and syllable. They change each tone and note and beat of their complaint and you cannot tune them out.

We went up to the sitting room to see her. The wooden case of the old Encyclopedia Britannica, the volumes inscribed to my father by his father, was stacked high now with dozens of books of poetry, biography, history, some of them my father's and some he had borrowed that month from his favorite library, the Providence Athenaeum. He did a lot of reading while he sat there with my mother. The pages of the new edition of his *Statistical Mechanics of Elasticity* lay scattered now in my brother's old room, on his desk and on his bed—all over the room, where my father had encamped to help with Mom. My mother was there in the sitting room in her chair by the air conditioner, screaming, or as my father put it, vocalizing. He went straight to her and gave her a loud buss on the cheek. I could tell he thought he was being tender, but the kiss looked and sounded angry. And then another smack. "So, Ponnie, how are you?" he said, and I could see how angry he was. He had just gotten a little taste of his good life, talking about books in the living room. Now back to those four Victorian wallpapered walls with his vocalizing wife.

For years, my father had tried to keep her illness as private as he could. Most of my mother's friends did not know that she was sick, and he discouraged those of them who did know from coming to the house. He worried that even the visits of our family disturbed her. When the phone rang—his friend Jan Tauc was calling from Brown to plan a lunch—I watched my father on the white phone in his armchair in the sitting room, hunching over and wrapping his fingers around the mouthpiece, trying not to let Jan hear his wife scream in the other armchair five feet away. Each "yes" started quietly and morphed into a rising wail that kept going in a horrific way, that long loud involuntary yes, like an acceptance that does not want to accept.

Yes! Yes! It was like the denial of every affirmation. It was almost worse than death itself.

I unpacked my suitcase in the guest room. Even through the closed doors of the sitting room, the guest room, and the guest bathroom, I could still hear her. When I finally walked back toward the sitting room, there was Dad coming out and closing the sitting-room door behind him. We faced each other at the top of the front stairs.

"I can't take this anymore," he said.

"What do you want to do?"

I looked at him. We were standing eye to eye on the landing. I had no idea what to expect. The only act that could help her now would be to do for her what I had once begged her not to do to herself. I was ready for anything.

"Let's go to the Athenaeum," my father said.

So we drove over to the Ath. It was closed. There was a hand-lettered white card on the door: "The library is closed due to excessive heat." My father returned a thin book of poems through the drop-off slot, and we trudged very slowly up College Hill to Brown's Rockefeller Library, the Rock. There, I looked up Lucretius in the computer catalogue, and we found him in the basement stacks on the B floor. We each chose a translation and sat in two carrels side by side underground, reading passages aloud to each other. The stacks were very cool, and deserted.

My father had picked a translation by a scholar from Providence College, where Ponnie had once worked as a reference librarian. He opened to a page at random and read something about "the atoms slashing into her cheek." What was the poet saying? We puzzled through it together. The atoms feel no pain, even if we do feel pain. He read the line again. Then he read on aloud, grinning with professional amusement to hear a voice from ancient Rome discussing atoms, laughing to hear a poem in which the idea of atoms sounds new and controversial—and, of course, laughing with cabin fever, with relief at getting out of the house. It amused him to hear Lucretius argue that atoms cannot feel anything, even though we human beings feel so much.

"Well," my father said, "I think we'd agree now."

{ FORTY }

First and Last

On the morning of September 11, 2001, the sky was very blue. I was sitting in my home in the Pennsylvania woods, reading a book called *Immortality: How Science Is Extending Your Life Span—And Changing the World,* by Ben Bova.

"You might be one of the immortals," the first page said. "Particularly if you are less than 50 years old, in reasonably good health, and live a moderate lifestyle, you may live for centuries or longer." I was forty-eight, so I just qualified.

While I was reading, I heard a knock at our front door. I heard my wife's voice answer it, and I heard the voice of the chimney sweep—he had come to clean the chimney, which had caught fire a week before. The chimney sweep's voice sounded agitated, but I kept reading. "Prophecy can be a tricky business, as many weather forecasters and stock market analysts can attest," Bova said. "Moreover, human nature tends to accept gloomy prophecies as probably correct, or nearly so, while optimistic prophecies are usually harder to accept and are often greeted with: 'That's too good to be true!' "

Bova had been an aeronautical engineer before he turned to writing science fiction and science fact, and he was positive that the challenge of immortality was nothing but a problem in engineering. "In an expression borrowed from the world of aviation, how far can we push the envelope?"

My wife found me and told me to come to the living room. I was surprised to see the chimney sweep standing in the middle of the room, watching our TV. The screen showed the image of a plane hitting a tower.

I put my book down. At the end of that day, I looked down and noticed the book still lying there on the coffee table.

At Rockefeller, my students' eyes had a naked look that fall. We all felt we were under siege, all under the sword together. And when the anthrax letters began arriving, it seemed likely to everyone in the country that the next great attack would be biological.

To us, that shock was almost as horrible as the first, though we should have known it all along. Biology could be as dangerous as physics, and genetic engineers could make cells worse than bombs. They could kill New York and leave the towers standing. The science of life was as charged with good and evil as the science of the atom had been in the century before. The whole subject, all our new hope and dread, seemed to float above the city in a dome of unknown dust. Some invisible hand might already have chosen death for us all. In the esplanade the very radiance of the morning, the mist between the crowns of the sycamores, might be particles of death.

In my office I could not sit on the sofa or read at my desk for more than a minute at a time. One line or two and I was back at the window. I was sick of the Path of the Explainers.

"Pro Bono Humani Generis, For the Good of Humankind," was the motto on the centennial banners that hung that year from the wrought-iron lampposts. The young scientists and doctors wore the same motto on the shoulder patches of their white lab coats as they hurried on the paths below my window, under the branches of the sycamores. The esplanade was also lined with sculptures that the Museum of Modern

Art had lent Rockefeller in honor of its centennial and the new millennium. Not one of the sculptures had a human form. Most of them looked like models of atoms and molecules, discreetly chained to the ground.

I had a private motto: "First and Last." I wanted what I had seen and felt when I was small to have some connection with what I would see, learn, and know in the end. I thought the whole human race wanted something like that. The beginning, middle, and end should make one unbroken story. The stem should lead up to the rim of a cup from which we could drink and still be ourselves.

"Who sings the song of the way things are?" Lucretius wrote in 55 B.C.E. "Who has the wings to soar on these discoveries?" I thought a thousand years must have passed since I had felt like singing.

Once I had longed for wings. Now I needed a new place to stand. Sometimes all I could do was stand at that window.

{ FORTY-ONE }

The Carriage House

"That's it?"

My wife Deborah had heard me describe the carriage house and she had expected something grand. Now she was surprised at how small and shabby the place looked. It was the summer of 2002. The carriage house was half dismantled and even the Victorian looked bare. The paint was peeling and it sat awkwardly on its plot of land, now that the carriage house behind it was almost gone. Most of the facade of the carriage house was stripped away and there were brown, bare plywood sheets nailed up on it.

We had driven up to Providence and then to Newtonville with my family in our green Volvo—our whole family, including Tinka, the dog.

We got out of the car. It was a day with a New England fall in it, a broken, bright, and mackerel sky, heavy white clouds with the sun breaking through again and again. Peggy Heywood sat by the carriage house, watching the two babies, Stephen's and Jamie's. Alex and Zoe were two and almost three. The babies sat on a pile of beams, eating purple Popsicles. Alex had fallen and cut his lip. It was bleeding. The Pops were to console him and chill his lip. A little red blood stained his Popsicle.

Our dog went bounding all over to greet everybody. The kids were excited. Slowly the dog settled down.

"The front of the carriage house needs to be rebuilt," Peggy said. "Because they found rotten beams. It had no known means of support. But all the walls are sound above a certain point. So it looks terrible, but whole areas won't need much work.

"The plan is, the ground floor will be Jamie's tools. The second floor will be Melinda's dance studio. She has a troupe, she could do choreographies. And the top could be an artists' and writers' studio.

"Someday," Melinda said, joining us from inside the house. "Stephen was optimistic about the time. I don't anticipate working in there soon. Maybe a year."

"There's a spider living in my wood," Zoe said.

Melinda said, "She's always going, 'my wood.' She thinks this is her wood."

Alex waved his Popsicle. "I throwed the spiders away."

"They are like twins," said Peggy. "They have a wonderful time together. And because Zoe is five months older, and girls are usually advanced compared to boys, he speaks incredibly well for two." He was always asking Peggy for his tools. "Here's your screwdriver," she told him the other day.

"That's not a screwdriver. That's a *chisel,*" he said.

"Yes, the tools are plastic, but that's not his preference," Peggy said. "In his favorite picture of himself, he's holding a drill and a mallet. The mallet [with a big rubber head] is the safest hammer we can give him. His favorite tool is a drill. He knows how to use it, so we keep the drill bit out."

Stephen's two young apprentice carpenters had disappeared inside the carriage house, and we could just hear a voice from inside, speaking in a kind of robotic monotone. There was something odd about the sound. It was almost a chant, "*Om Ah Hum.*"

"A voice enhancer," Melinda explained to us. "He has had it about a month."

When Stephen rolled up to us along the uneven ground in his wheelchair, his head lolled loosely and terribly from side to side. He and I went for a walk, Stephen rolling along through the streets at a gentle pace so that I could keep up. There was no traffic. I said I was still learning to talk with him. He stopped the chair in the middle of the street and looked up at me. Even with the voice enhancer, talking was hard: His face strained with effort and he looked me right in the eye. It was painful to watch him strain. I wanted to turn politely away, but if I did then I could not understand what he said. I had to keep my eyes locked on his and watch him carefully. This forced me to face what he was facing and hear what he was saying, and he was putting all he had into saying it.

"Try yes or no," he said, with grimaces of effort, staring into my eyes, straining with what looked like a great physical ordeal to make those muscles move and talk. He was pale, with his black hair short-cropped.

I asked him if he was still comfortable with my writing about him. He stopped the wheelchair again in the middle of Hyde Road and looked up at me. With our eyes locked like that, his face seemed to me to grow huge, gigantic, immense. But it was the effort that was immense—the intensity of the work. He seemed to swell until he was as large as all human effort.

"I'm fine," he said.

We rolled on. It took me a while to realize that when he spoke to me he was not stopping the chair in the street just for emphasis. Talking was such hard work for him now that he had to stop the chair to concentrate. When I asked him what it was like, he said, with the hint of a smile, "It's tiring."

A little later he stopped the wheelchair again, turned it to face me, and looked up.

"How's your mom?"

Stephen was as sensitive and observant as ever, concerned about everyone around him. He often thought his disease was harder on others than it was on himself. But he had learned not to say that aloud—for some reason, it made his mother fall to pieces.

Stephen was still Stephen. He was not about to get religious now, just because he was dying. The more he lost, the stronger he seemed to those who knew him well. He subscribed to what has been called the theory of normal accidents. On every construction job, there is a chance that one ridiculously small mistake—one number dropped from the blueprints, one bolt dropped from a beam—will start a crazy chain of events and bring down the whole house. That is what had happened somewhere inside his body. The same thing happened to about 5,000 Americans a year. His illness was a normal accident, and he refused to mythologize it.

Stephen did not mythologize Jamie's race, either. He did not lie awake at night counting over all the miracles that had helped his brother do what he was doing. Jamie's job at the Neurosciences Institute; his insane techie pride; their family pride, their being Heywoods together: Jamie had built amazing things from those beginnings, and Stephen was still convinced that his brother would find a treatment or a cure, maybe after Stephen himself was gone. But there was nothing supernatural about any of it. That is how big things always start. They have to grow from a couple of lucky little things. They are normal miracles.

Of course, Stephen was much too considerate to dwell on his atheism with Wendy, or his parents, or Jamie. His feelings were summed up in a little poem by Czeslaw Milosz, "If there is no God."

> If there is no God,
> Not everything is permitted to man.
> He is still his brother's keeper
> And is not permitted to sadden his brother
> By telling him there is no God.

Back at the carriage house, Stephen asked Peggy for lunch, and she took two cans from a pouch in the back of his wheelchair—Ensure, a liquid diet.

"Easy cooking," said Peggy.

She poured a can into his stomach tube, a pale chocolate-colored liquid going down the funnel and into the tube and right into his stomach. It was a strange sight, repellent and commonsensical. "Straight in," said Peggy.

"Mom, what do you want to do?"

"What do I want to do? I'm at your disposal."

"Where's the boy?"

"Where's the boy? Inside, with Melinda and Zoe."

Stephen said something I did not understand.

"You want to take a nap?" asked Peggy.

Stephen wheeled his chair into the carriage house and stopped on the center of a plywood square on the ground—the only flooring there was. The whole ramshackle place was over his head. He pressed buttons on the arm of his wheelchair, and the legs came up while the back went down, with quiet, competent whirs. He grinned toward me as he did it. He was showing off his gear.

Lying there in the middle of the carriage house, which gave more shade than shelter, reclining in his chair, Stephen looked so helpless that I hesitated to leave him. After a moment he opened his eyes and saw the expression on my face. Making the same giant effort as before, he said what he had always been saying. "I'm fine," Stephen said.

{ FORTY-TWO }

Throw Alex Higher

"Melinda took two beds. All the pictures, and all the bookcases, I notice," Jamie said. It sounded as if he had not been looking around much. It was September 2003. He would come home late from the foundation, flop down, and then run out the door in the morning. The Victorian was a mess. An old pal from Advent who had lost his job was living downstairs and buying the orange juice.

"There's a Circus Flora poster," I said.

"One left," Jamie said thinly. "There were two."

I noticed a big antique wardrobe of Melinda's, her costume bureau. Its two big doors were tied together with a sort of leopard sash or scarf. But the doors were ajar, and inside I saw empty hangers, with no show business left. On the walls there were still a few posters of Melinda or her mother belly dancing, posters with the type faces in Greek, Arabic, Cyrillic.

There were bare nails going all the way up the stairs. And at the head of the stairs, in a chaos of toys, a naked Zoe, who was at Jamie's place that night: "Daddy! Daddy! Where are you?"

"Yes, Zoe. Yes, Zoe."

"I need someone to come up with me," she cried fiercely.

Melinda had left for a small apartment in Waltham, where she hoped to restore herself and write. When he could, between crises at the foundation, Jamie was reading a book titled *Divorce Remedies,* which he liked for its problem-solving approach.

Jamie was having a hard time. The collapse of the technology bubble was long since complete. Thousands of computer engineers and genetic engineers were out of work, hundreds of foundations were going under. It seemed like the end of the end when Dolly died. She had a case of arthritis and a tumor in her lung. Her Scottish vets put her down on February 14, 2003.

That year, Jamie was on the road every week to fund-raise. He had now lost his money, his wife, and most of his old friends. Many of the young activists and idealists who had joined ALS-TDF in the start-up days were gone, too. Some of them Jamie had fired as he changed direction—he managed the foundation as aggressively as he drove a car. Others had quit because they thought he changed direction too often.

I could not quite forgive Jamie myself for the flat way he told me, one afternoon, about the fate of the project that had drawn me to his cause five years before. "The EAAT2 gene therapy is dead clinically," he said. His tone was all business. I think Jamie had put off letting me know for as long as he possibly could. He listed the reasons for his decision. The results from the lab tests of EAAT2 had not been strong enough to justify the risks of the operation. Washington was still hostile. For the time being, the whole field of gene therapy was a world of hope and disaster. Jamie claimed that in the end he had made a simple managerial call. "I'm very good at pulling the trigger."

Jamie had lost some hard battles, but he had also won a few. As the volunteers and activists left his foundation, he had hired people with experience in Big Pharma and biotech. He now ran the biggest animal drug-screening program for ALS in the world. That year, ALS-TDF was testing two drugs intensively. One was Ritonavir, a drug the FDA had approved to treat HIV, the virus that causes AIDS. The other was hydroxyurea, a drug approved by the FDA to treat sickle-cell anemia. Both drugs slowed ALS—slightly—in mice.

Jamie had expanded the course he called ALS 101. He taught it in

Boston and across the country, many weekends a year, to hundreds of patients and their families and friends. Many of the people who came to hear him started foundations of their own to raise money for Jamie. That year two professional golfers, Tom Watson and Jeff Julian, launched "Driving 4 Life" in honor of Watson's caddy, Bruce Edwards, who had ALS. The "4" stood for Lou Gehrig's uniform number. Proceeds would go to Jamie's foundation.

Jamie had now raised 12 million dollars. He was a leader in a growing movement of patient activists who were energizing the study of nerve death and regeneration. That year, the research seemed to be accelerating everywhere. Jeff Rothstein published the results of an important gene therapy experiment for ALS, and Matt During published another for Parkinson's. Rothstein was still working on EAAT2. So was Dave Poulsen, out in Montana. None of them thought they would cure the incurable overnight, but all of them were working hard.

Jamie put in twelve hours or more a day, seven days a week, and he had taken himself off salary again. In spite of all his longings for the house on the hill and the Aston Martin, he did the right thing. He agonized and agonized, and he did the right thing about the money.

When I asked Jamie if looking back he would do anything differently, he said, "If you ask me would I prioritize Melinda above those life-and-death decisions—no."

In the kitchen, a map was tacked on the wall, "Biotech Cluster in Greater Kendall Square," with the latest location of the foundation, below a faded color photocopy of the *New Yorker* photo of Jamie and Stephen. Melinda had tacked it up for Jamie before she left.

Jamie told me I was missing the point about the money. "If I could have figured out how to get rich, I would have," he said. "I mean the real question is, what's the best way to solve the problem. I think I did the right thing, but it's the right thing in terms of—what's the best way to go fast.

"I never had any qualms about getting rich."

When we said good-bye, Jamie gave me a stare and spoke of his hopes for the next drugs they would test at the foundation. That year he always closed our conversations with a positive note, and that long, dark, piercing stare: "Maybe we'll beat this thing yet." And this is how it works, this is how we do beat these things—with some people working in the box, and others way out of the box. But Jamie's foundation was broke. He was frantic to keep it alive. That year there was something terrible in his stare, as if he was disappearing into it, or felt himself disappearing. He saw none of his friends. Some weeks he did not even have time to see Stephen. He could hardly bear to look at Stephen now, and many people feared for him as if Jamie had become the dying man.

When I walked over to Stephen's cottage on Beverly Road, I found him reclining in the living room in his motorized chair with a Stephen Hawking keypad beneath his hand, and his hand lashed to it by an elastic. Alex sat on the couch with Wendy, watching cartoons. Wendy had a bad cold. The last time a cold went through the house, Stephen had come down with pneumonia and Jamie had taken him to the hospital in the middle of the night. Now he looked very cheerful. They all did. Wendy told me that just before I walked in he had said to her, by typing on his keypad:

I FEEL LIKE A KING.

"HEY," Stephen said to me. The robot voice of his computer was carefully chosen to deliver his sardonic zingers—his mordant one-liners—in the appropriate flourlishless flourish. It was late on a warm October afternoon, with warm sun, and the living room felt cozy, domestic, comfortable. He was enjoying his family, as comfortable at home as any young dad in a La-Z-Boy. In the living room that Stephen had designed with its cleverly angled skylights, the three of them were the soul of a young family at home.

"A blink means yes," Wendy said. "A slight shake of the head, no. He has a whole list of Alexander commands on the screen. And he can put them on continuous loop, too." He could repeat and repeat a command as needed.

Stephen called them up on his screen for me to read:

ALEXANDER NO VIDEO TILL YOU CLEAN UP.
ALEXANDER GET BACK IN BED.
ALEXANDER LIE DOWN UNDER THE COVERS.
ALEXANDER TURN ON THE LIGHT, AND GO PEE.
ALEXANDER NO MORE THROWING.
ALEXANDER NO MORE YELLING.
ALEXANDER BE CAREFUL.

"He can even read to Alex," Wendy said. "He's downloaded some books. Alex can sit in his lap, and Stephen can turn the pages of *Where the Wild Things Are* and read them aloud." The computer displayed each page and read each word aloud.

Alex asked Wendy if he could watch a video. Wendy said, "Well, let's ask your father, because he's trying to talk to his friend. Stephen, can Alex watch a video now?"

Stephen blinked yes, and Alex went to put in the tape. Wendy said to me, softly, speaking over Alex's head as he sat on the couch and watched, "He has no idea he's powerless."

"What do you mean?"

"If Stephen said no, Alex wouldn't argue. He would just throw himself back on the couch and go *Awww* and that would be that."

I told Stephen how much I liked the carriage house. It was almost finished now and it was quite beautiful. I waited awhile for him to type out his answer on the keypad.

IRONIC. BUILT THAT, WHILE MARRIAGE
COLLAPSED.

The flat voice suited his old, uninflected style of delivery. As I think of it now, it reminds me of that voice from the backseat as we were careering toward Matt During's lab. "It's us I'm looking out for."

"Feels like we're 'instant messaging,' " I said, and waited for his reply.

FEELS NATURAL.

I took a chance and told Stephen about a dream. In the dream, I had received a letter from someone who knew him well. I had opened the envelope and read the letter. It said there was something Stephen had not told me and wanted to tell me. Even as I launched into this speech, I could tell from certain sardonic hints in his face that I was getting too schmaltzy for Stephen. His fingers worked at the touch-pad. His stomach shook and a sort of grin tried to happen on his face, and his eyes literally shone. I guess I should have known better than to get so earnest, or ask for famous last words. His body did not fully cooperate, but his stomach was shaking, his mouth and face were working, his paralysis squelching a full-bodied laugh. Did my dream mean anything? I asked. Was there something he wanted to say? The robot voice spoke:

I'M PREGNANT.

Later on I told that story when I stopped in Providence on my way home. My mother was the first to laugh. So apparently my father was right—she really was still there. And when I told the two of them about Stephen's continuous loop: "ALEXANDER . . ." she laughed again. Over the years she had come to care about Stephen and Jamie, too. It was slightly awful to realize that she was still aware, inside her frozen face, but it was good to share another laugh.

In Newtonville, while Stephen and I talked, Alex leaned against my legs and watched the video. When his cartoon was over, I picked him up and flew him around the room, turned him upside down. "I'm keeping Alex." I really did want to take him home. There was another robot command:

THROW ALEX HIGHER!

At the door, I said good-bye to Wendy, and to Alex. Then I looked over and Stephen raised his eyebrows slightly by way of good-bye. Somehow it felt perfectly satisfactory, that good-bye, because I felt it was perfectly satisfactory to him.

Stephen reclined there in his automated chair with the late afternoon light streaming in his skylights, and I felt, he got where he wanted to get, he has what he wants, and he is there. That day it was Jamie I feared for. Stephen, impossible as it seemed, was happy and comfortable right where he was.

ACKNOWLEDGMENTS

I owe many people thanks, and the Heywood family most of all. They welcomed me into their lives during the kind of crisis that no family should have to face, but very many do. Special thanks to Melinda Marsh Heywood and Wendy Stacy Heywood for letting me read and quote from their journals. The book is done now, but I continue to learn from the Heywoods.

Ralph Greenspan, of the Neurosciences Institute, introduced me to Jamie Heywood and helped me follow the story as it grew. Arnie Levine, when he was president of Rockefeller University, invited me to explore the edge of medicine as writer in residence. The Alfred P. Sloan Foundation helped to fund my research for this book; many thanks once again to Doron Weber.

My agent, Kathy Robbins, saw the power of the Heywoods' story from the beginning. Kathy is superb with books and the whole world around them. At *The New Yorker,* Charles Michener edited my profile of the Heywood brothers, "Curing the Incurable." Thanks also to David Remnick, who faxed me encouraging words at the eleventh hour. At Ecco Press, Dan Halpern helped from start to finish with warm and wise advice. This book would not be what it is without his help.

My father encouraged me in the writing of this book in spite of his deep sense of privacy. He would much rather have kept our own story in the family, and I hope he will feel that the cause was good.

A few friends read the manuscript and made innumerable helpful comments: Andrea Barrett, Bobbie Bliss, John Bonner, Ralph Greenspan, Shirley Tilghman. My sons, Aaron and Benjamin, put up with my long hours, as always, and they made valuable suggestions, too. So did my extended family, and my friends in Bucks County.

At the Heywoods' family foundation, ALS-TDF, special thanks to Robert Bonazoli. At Princeton: Harold Shapiro, Peter Singer, Lee Silver, and the students of my seminars in science and literature. At Rockefeller: Betsy Hanson, Alice Lustig, and all my students there. At Arizona State University, where I spent a happy semester as a visiting professor, warm thanks to my friends Rick Creath and Jane Maienschein, and to our students. At Jefferson: Matt During and Paola Leone. At Johns Hopkins: Jeff Rothstein.

Special thanks also to Jacques Cohen and to Günter Blobel, who gave freely of their time and help.

Thanks to Gail Schmitt, who provided cheerful and competent help with research. Thanks also to the late Lynn Forbes, who transcribed many hours of taped interviews for this book from her home on Cape Cod. She is missed. And thanks to my cousin Jay Weiner and his wife Er-Dien. Their home in Santa Fe was a warm and beautiful place to finish this book.

I benefited from the reporting of many other writers. Special thanks to the science writers of the *Philadelphia Inquirer* for their powerful series of articles about Jesse Gelsinger's death.

In quoting from Lucretius, I used and freely adapted the work of many different translators, including Thomas Creech, Anthony M. Esolen, Rolfe Humphries, and W.H.D. Rouse.

For many kinds of help, I thank Steven Ascher and Jeanne Jordan, Roland Beckmann, David Colter, William Haseltine, Slawa Lamont, Sven Lindblad, Astrid Lueders, Alan and Monika Magee, David Magnus, Peter Mombaerts, Sanjay Nigam, Rabbi Sandy Parian, Tony Perry, Oliver Sacks, Thomas Schwartz, Kambiz Shekdar, Agata Smogorzewska, John Steele, LeRoy Walters, Michael West, and Steen Willadsen. I interviewed more than 100 scientists whose names do not appear in this book, and I thank them all. The book is richer for those conversations. Many thanks to the photographer Martin Schoeller for his remarkable pictures of Jamie and Stephen Heywood, and his generosity to the family. Christopher Potter of Fourth Estate read two drafts of the manuscript and made excellent suggestions.

I am grateful to everyone on the talented staffs of *The New Yorker,* Ecco Press, and the Robbins Office. At *The New Yorker,* special thanks to the fact checker Nandi Rodrigo, and to Charles Michener's assistant, Erica Youngren. At the Robbins Office, to Sarah D'Imperio, David Halpern, Sandy Bontemps Hodgman, Rick Pappas, and Teri Tobias. At Ecco, to Jill Bernstein and Robert Grover. The book was lucky to have Claire Vaccaro as its designer, Lyman Lyons as its copy editor, and Mareike Paessler as its senior production editor. David High and Ralph del Pozzo of High Design designed the perfect cover.

And I thank my wife, Deborah Heiligman, the closest reader of each draft, and each book, first and last.